*Toward Unity among
Environmentalists*

Toward Unity among Environmentalists

BRYAN G. NORTON

Professor of Philosophy
Georgia Institute of Technology

New York Oxford
OXFORD UNIVERSITY PRESS

Oxford University Press

Oxford New York Toronto
Delhi Bombay Calcutta Madras Karachi
Kuala Lumpur Singapore Hong Kong Tokyo
Nairobi Dar es Salaam Cape Town
Melbourne Auckland Madrid

and associated companies in
Berlin Ibadan

Copyright © 1991 by Oxford University Press, Inc.

First published in 1991 by Oxford University Press
198 Madison Avenue, New York, New York 10016-4314

First issued as an Oxford University Press paperback in 1994

Oxford is a registered trademark of Oxford University Press

Library of Congress Cataloging-in-Publication Data
Norton, Bryan G.
Toward unity among environmentalists /
Bryan G. Norton.
p. cm. Includes index.
ISBN 0-19-506112-8
ISBN 0-19-509397-6 (pbk)
1. Environmental policy—United States—
Citizen participation—History.
I. Title.
HC110.E5N67 1991 363.7'058'0973—dc20 91-16719

3 5 7 9 8 6 4 2

Printed in the United States of America
on acid-free, recycled paper

For A. A. J.

Preface

Lewis Carroll told of a map—a magnificent map—of the German country-side. Its scale was one inch to one inch. Every detail was represented. Alas, however, it could not be used; when unfolded, it shaded the crops.

In this book I have tried to provide a map of the countryside of environmentalism in the United States, and I have tried consciously to err in the other direction—to look not at trees but at forests, to subtract details in the hope of seeing larger patterns, and to look at environmentalism as a movement in larger resolution than would be possible within the confines of a single discipline.

When Resources for the Future graciously invited me to be Gilbert White Fellow for the year 1985–1986—the first philosopher to work at RFF—I told them I would do research on the "values and objectives" of environmentalism. I'm sure that seemed a bit nebulous to them, and I am extremely grateful for their openness to and support for an approach very different from their own. They gave me an office and a word processor, and left me to wander hallways lined with open doors, most of them occupied by hard-nosed resource analysts who were willing to talk, although they no doubt thought my project a bit frivolous.

A couple of weeks into the project, I was still feeling my way; I visited a few libraries at the large environmental organizations that have headquarters in the Washington, D.C., area, reading organizational statements of purpose and charters and generally poking around. At one of the organizations, the kindly librarian—who had worked there for thirty years and had seen many researchers come and go—asked me to describe my topic so that she could help me to find relevant materials. I said something about studying the goals of environmentalists, what they are trying to accomplish. She looked over her glasses and said, "Young man, don't you think you should limit your topic?" It was the best advice I'd had since I got the same suggestion from my twelfth-grade English teacher.

Not given to accepting good advice, however, I persisted, knowing all along the many pitfalls of interdisciplinary work. The curse of such work, it has been said, is that one learns less and less about more and more until one knows nothing about everything. I have tried to stop just short of that limit, to subtract elements like a sculptor until I began to see large patterns emerging from the amorphous mass that is the modern environmental movement.

I was driven, throughout the enterprise, by an initially inarticulate sense that, while we know a great deal about environmental problems and a little about how to solve them, the large-scale understanding of the field—the generally accepted assumptions about what environmentalists are up to— are off the mark. In ten years of studying environmental ethics and hanging around on the fringes of the environmental policy community, I had gradually become aware that the apparently important value questions discussed and formulated by philosophers did not bear very directly on the concerns of environmental activists, and that where environmental ethicists saw distinctions and dichotomies—a black-and-white world of good and evil— environmental activists saw a world of grays and lesser evils.

As the resource economist Alan Randall once said, "Resource analysts live high on the information food chain." I have, in this sense, lived as a top-level predator, and I have benefited greatly from the recent spate of excellent books and synoptic articles in the fields of environmental sociology, politics, conservation biology, environmental history and biography, environmental management, and environmental ethics, as well as the less academic writings of American naturalists and activist environmentalists. I have also tried to read as many primary sources as possible within various fields, although a careful look at the endnotes will reveal that I have relied heavily on synoptic works in many cases.

The most difficult problem I faced, however, was the lack of an adequate and consensually accepted vocabulary, either colloquial or technical, that could be employed without distortion across disciplinary boundaries. Controversial theoretical assumptions and debatable value preconceptions taint much of our language in the field; I realized that the various participants in environmental debates see events so differently, and describe them in such diverse vocabularies, that my cartographic efforts lacked the fixed points that are afforded by shared theoretical principles and accepted linguistic conventions.

Therefore, I had to develop a vocabulary as I went along and introduced the imprecise, but I think helpful, method of "worldview analysis," a method recognizing that our ideas, concepts, and values tend to exist in associated clusters rather than as interchangeable atoms; it recognizes also that the meanings of those clusters are intimately related to their context in concrete activities. The enterprise, which I have only begun here, of developing a framework of concepts that is loose enough to be applied across disciplines and yet tight enough to retain some meaning in its various applications, remains an urgent task; until it is accomplished, environmentalists and their supporters in the natural and social sciences will continue to speak different languages. A common interdisciplinary language is the first step toward a unified theory of environmental stewardship. I hope that my efforts stimulate further efforts to develop an adequate vocabulary for discussing environmental problems in their many forms.

In spite of the many obstacles, I believe that a look at the larger picture is at this time essential. Environmental problems are resurging into public consciousness. And, while environmentalists have no shortage of "solutions" for particular problems, they have brought forward for public discussion no clearly articulated goal, no vision for the future, that could serve to unite a broad political movement to address the general depletion of resources and decay of our environmental context.

What do environmentalists want, anyway? Is environmentalism a "movement" or a collection of interest groups hiding under a single umbrella-like label? If one listens to what environmentalists *say*, one hears a cacophony of programs and explanations. Some environmentalists sound like economists, speaking a language in which values are measured in present dollars. Others say they are protecting resources because of obligations to future generations, a concept not expressible in mainline microeconomics. Still others call themselves "Deep" Ecologists and insist that nature has value in its own right; to violate nature is to them a moral evil. Yet others feel comfortable with a Judaeo-Christian vernacular and speak of Christian stewardship and obligations to God to respect the garden that he created for us.

This babel of voices leaves environmentalists ill equipped to build a unified conception of environmental management, and it hampers them in communications with the general public. How can they propose a unified and integrated conception of environmental management, a blueprint for living in harmony in nature, if they cannot accept a common language in which to express it?

Some readers will question my enterprise, which assumes that there is *an* environmental movement. Why assume "environmentalism" is more othan a label for diverse interest groups, each seeking its egocentric agenda? For such readers, I can only ask that, before giving up, they examine the map I have drawn of the larger territory. I have concluded that we have not seen the unity of environmentalism because we have been looking in the wrong place. I have, in my search, tried not to use environmentalists' rhetoric—the explanations they give for what they do—but their actions—the policies they actually pursue—as the fixed points on my map.

My approach, therefore, is broadly pragmatic. I have tried, however, never to use my philosophical pragmatism as a premise, but only as a constant guide. I see pragmatism not as a set of metaphysical principles, but as a method. That method starts with the idea that, in some important sense, actions are more basic than words, and that words get their meanings from the actions with which they are associated. Concepts are thus seen as tools of understanding, but tools that function best in the context of action. I hope, therefore, that this book displays a healthy skepticism for justifications based upon principles that are known intuitively, or a priori—true independent of empirical experience.

Accordingly, the book is also skeptical of grand dichotomies, which are so often the fellow-travelers of intuitive, a priori knowledge. Grand dichotomies, as John Dewey recognized at the turn of this century, thrive only in ivory towers; when held up against the real world, they do not fit, and are tumbled about and scratched. Underneath, one usually finds a continuum with an oversimplification superimposed.

My goal has been no less than to propose an integrated theory of environmental management, which requires as a component an understanding of the human role in natural events. If I have failed to present a map that is correct in all of its details, I shall not be surprised. My hope is only that my ambitious, even foolhardy, explorations will stimulate others, more capable than I, to explore the largely uncharted territory of environmental action and value and to carry forward the search for an integrated conception of environmental protectionism.

I owe thanks to numerous persons who provided important advice and reactions while I was working on this project, especially Peter Brown, Terry Davies, David Ehrenfeldt, Mark Sagoff, Thomas Lovejoy, and Edith Brown Weiss. Stephen Petranek, of the *Washington Post*, sharpened my journalistic skills by patiently helping me with the original version of the sand dollar story. As the manuscript took shape, chapters were reviewed by Jim Banner, J. Baird Callicott, Stanley Carpenter, and Riley Dunlap. I have already mentioned the essential support from Resources for the Future, and I am equally indebted to individuals there—John Ahearne, Jim Banner, Allen Kneese, Danny Lee, Paul Portney, Norman Rosenberg, Clifford

Russell, Pierre Crosson, and especially Robert Mitchell—for countless helpful conversations about my topic.

It is with pleasure that I acknowledge my debt to the environmentalists, policy experts, and agency personnel who agreed to be interviewed on their views of environmentalism today. While few direct quotations from the interview tapes have found their way into the final text, the wisdom I gleaned from the interviews shaped my general approach and influenced my stance on every page. In particular, I thank Richard Ayers, Michael Bean, David Brower, William Butler, Terry Davies, Laurence Jahn, Charles Little, Jack Lorenz, John McGuire, Gaylord Nelson, Paul Portney, Neil Sampson, John Seiberling, William Turnage, and Michael Wright for long and informative interviews.

After four years of thinking and writing I ended up with a comprehensive but unwieldy manuscript. Like so many authors, I was unable to jettison large portions of my work even as I realized line-by-line editing would never achieve the goal of a concise and readable book. Sara Ebenreck rescued both the manuscript and my fragile sensibilities by performing major surgery under the anesthetic of kindness and encouragement. I am indebted, also, to Richard Wallace, who spent countless hours improving the manuscript in many major and minor ways, checking quotations, and tracking down obscure references. The editors at Oxford University Press—initially William Curtis, and later Paul Schlotthauer—have provided helpful guidance and advice throughout the project. Shirley Schultz and Norma Singleton assisted in preparing versions of the manuscript.

Finally, because elements of the manuscript have been published elsewhere, I wish to acknowledge the willingness of various publishers and editors to allow me to integrate material from their publications into the current text. Sections of Chapter 1 were written for the *Washington Post Sunday Magazine;* portions of Chapters 3 and 8 appeared in *Beyond the Large Farm,* edited by Paul Thompson and Bill Stout and published by Westview Press, and also in my essay, "Context and Hierarchy in Aldo Leopold's Theory of Environmental Management," which appeared in *Ecological Economics.* That same journal also published some materials that appear in Chapters 5, 11, and 12 as "Intergenerational Equity and Environmental Decisions." Finally, some portions of Chapter 9 were prepared for publication in *Wilderness* and were adapted for use in this book.

Atlanta   B. N.
February 1991

Contents

Contents

Foreword

In 1983 the Center for Philosophy and Public Policy embarked on a chal-
lenging project to look for common ground among philosophers and envi-
ronmental/natural scientists. The first meetings were necessarily rocky as
we struggled to understand one another's language and look at the various
environmental and human dilemmas from our differing perspectives. Although philosophical concerns had not been excluded by environmentalists, figures like Muir, Pinchot, or Leopold, however passionate, generally
had been attracted first by nature and science and only later became more
contemplative about the philosophical ramifications. In this instance there
was an active attempt to create a dialogue among disciplines. It was the
beginning of an effort, which finds its culmination in this volume, to define
the philosophical basis of what Thomas H. Huxley so aptly termed "man's
place in nature."

The field of environment speaks in many tongues: technical, lay, scientific, aesthetic, moral, privileged, disadvantaged, rational, empassioned. Yet
it is—or should be—a topic of concern to all, whether they be Chesapeake
Bay watermen, Kayapo Indians, or those concerned about environmental
disruption on a global scale and the loss of tens of thousands of species
annually. The environmental handwriting *is* on the wall. But will we collapse into an ineffective cacophonous babble or find a unity that can draw
on the strength of our diversity? Will we become a society "dead of its own
too much," to paraphrase Aldo Leopold, or will we find a way to come to
terms with our constrained freedom?

Bryan Norton can see the world in a sand dollar, and has a journalist's talent for lighting the way through the complex and arcane. He believes, and I am convinced shows in this volume, that there is a fundamental unity in environmentalism. The conservationist (*sensu* Pinchot) who believes our natural resources are for our use, today believes that use must constitute a matrix within which preservation occurs. In other words, use can mean non-use in a non-consumptive sense. Technical quibbles aside, a conservationist can see that the current raging controversy surrounding old-growth forest in our Pacific Northwest is much more than the media's simplistic "spotted owl vs. jobs"; that the endangerment of the owls is merely symptomatic of an industry based on a vanishing resource—and who could defend the logic of using the last crumbs of a resource when there are so many other values involved in retaining them? And any preservationist, however militant about maintaining wild places, knows full well that their protection rests on a stable human population/resources equation elsewhere.

The danger is that in struggling for the conservation resources to redress the imbalance between society and the natural world, we will be less effective if we fail to recognize this fundamental unity and seek synergies to strengthen the effort. Worse, we could fail so completely that the imbalance will tilt further until society is paralyzed by the selfish scramble for dwindling natural resources. To prevent this is the challenge. Surely our conscious nature will allow us to meet it.

The illiterate Pennsylvania farmer John Bartram was impelled by simple contemplation of a wildflower to become the greatest American botanist of colonial times. So, too, almost all the great figures in this book show a capacity to draw inspiration and strength from the world of nature. I believe that we will not be able to make peace with the natural world until we value our fellow species as we value our own works of art. The recognition of this unity of value and inspiration must expand beyond environmentalists to include all of human society. If that can be achieved, we will receive the ultimate gift from our environment that preservationist and conservationalist, shallow ecologist and deep ecologist, will celebrate alike.

Thomas E. Lovejoy

*Toward Unity among
Environmentalists*

1

The Environmentalists' Dilemma

Dollars and Sand Dollars

The poignancy of the dilemma facing advocates of environmental protection was dramatized for me in an encounter with a little girl. It was a sleepy, summer-beach Saturday and I was walking on a sandbar just off my favorite remnant of unspoiled beach on the north tip of Longboat Key, Florida. The little girl clambered up the ledge onto the sandbar, trying not to lose a dozen fresh sand dollars she cradled against her pushed-out and Dan-skinned stomach. I guessed she was about eight.

Thirty yards away, in knee-deep water, her mother and older sister were strip mining sand dollars—they walked back and forth through the colony, systematically scuffing their feet just under the soft sand on the bottom of the lagoon and bending over to retrieve each disk as it was dislodged. Their treasure was held until collected by the eight-year-old transporter, whose feet were too small to serve as plowshares. Gathering the sand dollars at the point of excavation, she relayed them to the sand bar where a considerable pile was accumulating near the family's beached powerboat.

Many months earlier, I had noted how the fickle current through Longboat Inlet had begun to dump sand in a large crescent spit out into the Gulf of Mexico, forming a waist-deep lagoon. Next came a profusion of shore birds and the colony of sand dollars that multiplied in the protected water, and then came the little girl and her family in their powerboat.

I was startled by the level of industrial organization; even the little girl executed her task with square-jawed efficiency. I engaged her as she emerged onto the sand bar. "You know, they're alive," I said.

3

"We can put 'em in Clorox at home and they'll turn white."

I asked whether they needed so many. She said, "My Momma makes 'em outta things."

I persisted: "How many does she need to make things?"

"We can get a nickel apiece for the extras at the craft store." I sighed and walked away. Our brief conversation had ended in ideological impasse.

But I was troubled. How could my indignation be stilled so simply? Must the environmental conscience always give way to economic arguments? As I wandered off, I analyzed the short and unsatisfactory debate. I had begun by expressing my concern for life, for the several hundred green discs drying in the afternoon sun. Yet I'd have felt silly saying, "Put them back, they have a right to live." I'd have felt silly because I don't think it's immoral for little girls to take a few sand dollars from the beach, any more than I had been immoral when I had red snapper for lunch that same day. I felt ill-equipped to make my point, about which I had little doubt, that the little girl should put most of the sand dollars back. If I admitted that sand dollars are just resources, like chunks of coal, salable in an available market, I could not at the same time argue that the little girl should put most of them back. Once sand dollars are economic resources, their value is counted in nickels. Therefore, I could not express my indignation in the language of economic aggregation.

Nor could I precisely express it in the language of rights of sand dollars, especially not if that language is given its accepted meaning in the tradition of John Locke and Thomas Jefferson. I did not find it self-evident that all sand dollars are created equal with little girls. Imagine instead that I had encountered the little girl with a half-dozen sand dollars submerged in sea-water in her bucket and she had said, "We're going to cut up a couple and put the rest in our saltwater aquarium and watch 'em." If I appeal to the rights of sand dollars as individuals (or, even somewhat more weakly, to the intrinsic value of individual sand dollars), I would have to object to this purely instrumental use of sand dollars in a rudimentary science lesson.

I faced the environmentalists' dilemma;[1] it was a dilemma, not because I did not know what I wanted the little girl to do, but because I could not coherently explain *why* she should put most of them back. If I chose the language of economic aggregation, I would have to say she could take as many as she could use, up to the sustainable yield of the population. On this approach, more is better—the value of sand dollars is their market value, and I could not use this language to express the moral indignation I felt at the family's strip mining sand dollars and hauling them away in their powerboat. To apply, on the other hand, the language of moralism, I would have to decry the treatment of sand dollars as mere resources; I would have to insist that the little girl put *all* of them back. Neither language could express my indignation *and* my commonsense feeling that, while it was not wrong for the little girl to take a few sand dollars, she should put most of them back—the aggregationist approach to valuing sand dollars would prove too little, and the moral approach would prove too much.

Consider again the altered scenario in which the little girl takes a half-dozen home in her bucket to be cut up or imprisoned in an aquarium. Suppose the little girl takes them home, and they are, predictably, dead in a week, but that the little girl attains an interest in biology, eventually becoming a marine biologist who works to protect echinoids. If sand dollars had myths and legends, the sacrificed sand dollars might be worshiped as saviors of their kind. And to the little girl, also, they would then have been far more valuable than nickels. It is this sense of respect for sand dollars as living creatures, worth more than mere nickels but less than little, round people, that I could not express in either the strict language of moralism or in the language of simple economic aggregationism. I knew I wanted to get the little girl to put most of them back, and to respect the remaining ones as living creatures from whom we might learn something worthwhile; I was torn between two inadequate languages for expressing the value of sand dollars. In this sense, the environmentalists' dilemma is primarily a dilemma in values, conceptualizations, and worldviews more than a dilemma regarding actions and policies. It affects mainly how environmentalists explain and justify their policies, and only occasionally and tangentially does it affect those policies themselves.

My conversation on the beach represents, in microcosm, a larger dilemma facing environmentalists. I know that this practical and industrious family would not be moved by speeches for sand dollar liberation, however eloquent. That argument had been cut short, rendered irrelevant by the little girl's utilitarian reply. Sand dollars are by no means an endangered species, so that line of argument wasn't applicable.

Once I'd given up my moral high ground and asked only whether they needed so many, I'd conceded the utilitarian value of sand dollars. If a few are useful as commodities, surely more are correspondingly so. Of course I could have given her the conservationist line, that she should take only the sustainable yield of the colony. But I didn't have the faintest idea how to do a population model to show the little girl that she'd exceeded permissible levels of exploitation and, even if I could have, it wouldn't have satisfied me. I wanted to say more. So I fell silent, stymied.

As in my conversation with the little girl, environmentalists often begin by implying that there is something morally wrong in the systematic exploitation of nature, something that cannot be fully expressed in the language of scientific resource management and maximum sustainable yields. When the heat is on, however, they retreat to the solid ground of economic arguments, as I did when I tried the "How-many-do-you-need?" routine.

Environmentalists face two crises, one external and one internal. Against outsiders, they must continually defend their hard-won successes and urge new reforms against advocates of commercial interests who insist that environmental legislation ought never to disrupt "economic efficiency." Examples abound. The Reagan Administration set out, almost immediately after taking office, to invalidate all regulations, environmental and otherwise, that could not be shown, through a benefit-cost analysis, to promote economic efficiency. After decades of trying, environmentalists intent upon saving the Chesapeake Bay from progressive deterioration of water quality caused by industrial dumping and run-off from farmers' fields and subdivision yards achieved a regional plan for protecting the bay. Now they are fighting innumerable battles on a local level as development interests in individual communities pressure local governments to implement the plan by fleshing it out with maximally lenient local land-use plans.

While these external challenges command the attention of environmentalists, a theoretical crisis, in language and worldviews, causes paralysis and miscommunication within the movement: There has emerged within the movement no single, coherent consensus regarding positive values, no widely shared vision of a future and better world in which human populations live in harmony with the natural world they inhabit.

The environmentalists' dilemma, which is primarily a dilemma in ultimate values, results in inarticulation when environmentalists discuss, explain, and justify their policies. To the extent that utilitarian and more preservationist approaches are seen as exclusive choices—as *opposed* rather than complementary values—it follows that I must choose between two inadequate languages to express my indignation. Neither the language of biocentric moralism nor the language of utilitarianism was adequate to explain and justify my view that the little girl should put most of the sand dollars back.

Historically, it has been useful to speak of two divisions of the environmental movement, "conservationists" and "preservationists," because some environmentalists have faced this dilemma squarely and have opted for one horn of it or the other. Most conservationists see natural ecosystems and other species as resources and are concerned mainly with the wise use of them. Finding its philosophical roots in the ideas of Gifford Pinchot, first official forester of the United States, this group judges all questions according to the criterion of the greatest good for the greatest number in the long run. The members of this faction, who are often trained as professional resource managers, have usually exerted their influence through control of governmental agencies such as the Forest Service and the Bureau of Reclamation. These environmentalists apparently diverge from the value system of their more commercially concerned opponents in industry only in insisting that costs and benefits of development and exploitative projects be computed over longer frames of time. Conservationism, or wise-use environmentalism, emphasizes avoidance of waste in the present pursuit of economic well-being. Thus, while natural ecosystems and other species are resources to be used wisely, they are very definitely to be *used* for human purposes. Pinchot once said, "The first great fact about conservation is that it stands for development. . . . [Its] first principle is the use of the natural resources now existing on this continent for the benefit of the people who live here now."[2]

Conservationists, especially those who are trained in resource management and those who work in government resource agencies, have generally applied concepts and a value system that tend toward economic reductionism, which interprets values as individual preferences expressed in free markets. The value of a sand dollar, on this view, is what someone is willing to pay for it. This reductionistic approach has led to a long-standing collaboration of conservationists with economists and to a tendency to pose questions in quantified terms in which information on resource use and its consequences can be aggregated and presented in dollar terms.

Opposed to this group is another, often called "preservationists," which is committed to protecting large areas of the landscape from alteration. This faction derives its spirit and mandate from John Muir, who was the first president of the Sierra Club (in 1892). Muir saw his quest to preserve nature as a moral one. He rejected or reinterpreted the Christian views of monotheism and the Judaeo-Christian idea that nature exists for the sake of humans, arguing that the dogma "that the world was made especially for the uses of men" was the fundamental error of the age, and that "Every animal, plant, and crystal controverts it in the plainest terms." Muir railed against human arrogance that judges nature only according to human values:

> How narrow we selfish, conceited creatures are in our sympathies! how blind to the rights of our fellow mortals! Though alligators, snakes, etc., naturally repel us, they are not mysterious evils. They . . . are part of God's family, unfallen, undepraved, and cared for with the same species of tenderness and love as is bestowed on angels in heaven or saints on earth.[3]

Initially, Pinchot, Muir, and their disparate groups of followers worked together in opposition to the timber barons and other wasteful exploiters of natural resources. Both of these leaders, especially Pinchot, can today be thanked for creating the immense National Forest system. But Muir and Pinchot quarreled over grazing in the national forest preserves, and opposed each other bitterly over the plan to dam Hetch Hetchy, a beautiful canyon in Yosemite National Park. Pinchot allied himself with the developers: "As to my attitude regarding the proposed use of Hetch Hetchy by the city of San Francisco . . . I am fully persuaded that . . . the injury . . . [caused] by substituting a lake for the present swampy floor of the valley . . . is altogether unimportant compared with the benefits to be derived from its use as a reservoir."[4]

Muir stated the case for preservation: "These temple destroyers, devotees of ravaging commercialism, seem to have a perfect contempt for Nature, and instead of lifting their eyes to the God of the mountains, lift them to the Almighty Dollar."[5] "Dam Hetch Hetchy! As well dam for watertanks the people's cathedrals and churches, for no holier temple has ever been consecrated by the heart of man."[6] Muir's pantheism implied that humans exist as part of a great spiritual whole. We worship that whole, the creator and sustainer of us all (which Muir identified with nature itself), he thought, by preserving and studying the most spectacular and beautiful areas as shrines. But Muir's heretical theological reasoning was never made explicit in his public writings. Indeed, he was referred to as a "man of God" by his contemporaries, and he appealed effectively to the powerful tradition in American protestantism, traceable to Johnathan Edwards, that saw nature as God's messenger to humans.

With scientifically trained professional conservationists lined up against Muir over Hetch Hetchy, he appealed to the public. In reviewing the revised edition of Muir's *Our National Parks* in 1909, the *New York Times* declared: "It is the sentimentalist like Mr. Muir who will rouse the people rather than the materialist."[7] And rouse them he did. Against all odds, Muir and his band of amateur preservationists held up the Hetch Hetchy Project for more than a decade. But when Woodrow Wilson took office and swung his weight in favor of the dam, the bill was forced through Congress by a narrow margin. The despondent Muir died shortly thereafter.[8] But his flaming rhetoric had created a powerful force of moralism in American environmentalism. That force has, from time to time, come to the fore as a political power, as when the Sierra Club, under the radicalized leadership of David Brower in the 1950s, succeeded in quashing a proposed dam in Dinosaur National Monument.

When I asked environmentalists what they thought was meant by the terms "conservationist" and "preservationist" today, most of them said the terms were meaningless or that the old terms have deteriorated into pejoratives, with "preservationist" being an epithet for someone who wants to "lock up" resources, and "conservationist" referring to a person who has demonstrated too great a willingness to compromise with economic interests. Since there seems to be a real question whether these terms are meaningful and useful today, I will use them sparingly and usually in historical contexts.

For the purposes of this book, however, I think it best to question the continuing usefulness of categorizing environmentalists as *two exclusive groups.* Here, Economic Aggregators and Moralists will mainly appear as ideal arguers. They are idealized spokespersons who have "bought into" one worldview or the other. A worldview will be considered as a constellation of concepts, values, and axioms that shape the world its proponents encounter. We can count on idealized spokespersons to give the best account they can, with their respective worldviews, of any given situation. It is best to think of Aggregators and Moralists in this way, as idealized arguers, because we can then leave open the question of whether most environmentalists, in their day-to-day activities, exclusively employ one or the other of these worldviews. Using this ploy, we can allow real individuals to be the spokespersons for Aggregators or Moralists in specific situations, without presupposing that environmentalists are always and permanently arrayed in two exclusive camps.

Muir's and Pinchot's respective successors in the modern environmental movement have more recently cooperated by maintaining an uneasy coalition. In general the environmental movement achieves its greatest unity when confronted with a sustained attack on environmental policies and programs, such as the one mounted by the President and his Secretary of the Interior James Watt in the early years of the Reagan presidency. The divergent elements also unify to defend specific resources and natural areas against threats of environmental degradation. In spite of these broad agreements on policy, however, the environmental movement still faces a dilemma: There has emerged within the movement no shared, positive understanding of the human relationship to the natural world; consequently, environmentalists lack a consensually accepted set of ideals and values. They therefore ricochet back and forth between two apparently exclusive worldviews and sets of value assumptions.

THE CHOICE between the legacies of Muir and Pinchot also presents itself to the environmentalist as a political dilemma: To follow Muir and grant rights to rattlesnakes is to embrace a radical ideal, one that appeals deeply to a small but committed minority that rejects the thoroughgoing anthropocentrism of our Judaeo-Christian tradition. This ideal, which elevates all nature to moral standing, calls into question the very idea motivating the American faith in Adam Smith's invisible hand, the idea that the path of economic development should be guided by a free market. Since nature has no dollars to spend, its voice cannot be heard in a marketplace; on any easily intelligible theory of the rights of rattlesnakes, these rights will limit the free choices of industrialists and consumers to buy and sell, to exploit and make profits. Embracing rights for rattlesnakes therefore damns the environmentalists, at least until there are fundamental changes in the value system of mainstream American society, to appealing to a very small audience of quacks and cranks, who are out of step with the economic values of our period of history.

But to follow Pinchot, to forget Muir's impassioned moral rhetoric, reduces environmentalists to a role as one more interest group, fighting for clean air, for clean water, for protection of the National Parks. These activities appear, politically, as no more than spirited support for strongly felt preferences. Clean air must be "balanced" against jobs and economic growth, and if consumers want clean air, they must be willing to pay for it in forgone jobs and dividends. On this side of the dilemma, environmentalists have lots of company. Everywhere there are interest groups shouting to protect their piece of a limited economic pie, and environmentalists are in danger of being entirely drowned out in the frantic melee, as everyone from profiteers to moral zealots attempts to focus governmental resources on social problems both real and imagined.

The environmentalists' dilemma, then, manifests itself in a number of ways. Among those who have opted for one or the other horn of the dilemma, it manifests itself in factionalism and distrust of those perceived to have joined the other camp. Other environmentalists remain uncommitted and uneasily embody both factions as internal personnae. The resulting theoretical schizophrenia can paralyze us with inarticulation and humble us in a debate with an eight-year-old in the sand dollar business.

The dilemma is especially evident in accounts of, and commentaries on, the progress of environmentalism. Historians, social scientists, and philosophers who have discussed the movement have been quick to see dichotomies and polarities. For example, the historian Stephen Fox emphasizes that Pinchot derived his strength from professionals, scientific forest managers and bureaucrats who made their living in exploiting or regulating resource use, while Muir drew upon the enthusiasm of amateurs motivated by an almost-religious zeal for the preservation of nature. Political scientist Lester Milbrath notes that environmentalism is a value-oriented reform movement but insists that "we must make a distinction between environmentalists who wish to retain the present socio-economic-political system and those who wish to drastically change it."[9] Philosophers who have discussed environmental values have concentrated almost exclusively on the dichotomy between anthropocentric (human-related) values and biocentric (nature-oriented) values.[10]

While these dichotomies do not all draw precisely the same distinction, they emphasize the polarization of environmentalists and suggest that the polarization derives from essential differences regarding values. For better or worse, these diverse but related dichotomies were given a generic characterization by Arne Naess when he distinguished a "shallow" from a "deep" ecology movement.[11] Naess's categories generally serve to characterize clusters of individuals who largely fit the Pinchot/conservationist mold and the Muir/preservationist mold, respectively. We must leave it an open question whether this dichotomy corresponds directly to our separation of Moralists and Aggregators. When pressed for an essential difference marking this generic distinction, Naess and his followers emphasize that shallow ecologists retain the anthropocentric view that the natural world exists as resources for the use of humans, while deep ecologists adopt the biocentric view that nature, as well as man, has intrinsic value and that it should be preserved for its own sake.

While Naess's provocative and tendentious characterization of conservationists as "shallow" environmentalists represents an extreme example, it is generally true that academic and social commentary on the environmental movement has accepted and even reinforced the dilemma and the deep polarities it evokes. Historical and sociological accounts that emphasize the different training and backgrounds of conservationists and preservationists, as well as philosophical analyses that concentrate on the dichotomy between anthropocentric and biocentric value systems, conspire to reinforce these polarities. This emphasis on deep underlying differences in values forces us to wonder whether the environmental movement is a "movement" after all. If the individuals and groups often referred to as "environmentalists" embrace no common values, then why assume that the environmental movement has a true and lasting identity? If left unchallenged, these suspicions undermine the task at hand—to understand a movement. Presumably, it is a movement *toward something*. To emphasize only the disparity of visions pursued by the various contributing factions is, in effect, to deny that environmentalism is a movement at all.

The purpose of this book is to challenge the suggestion that environmentalists hold no common ground, and the associated suggestion that environmentalists represent at best a shifting coalition of interest groups. That suggestion is implicit in the persistent emphasis, among historians and commentators, on the competing worldviews and value frameworks that constitute the vocabularies in which environmentalists argue their political case. According to the thesis of this book, those who see only chaos and confusion, internal disputes and dissensions, and those who deny that environmentalism is a unified social movement, are looking in the wrong place for the unity of environmentalism. Environmentalists, I am admitting at the outset, have not accepted a common and shared worldview, and those who look for unity in the explanations and rhetoric of environmentalists will be disappointed.

I will pursue a different strategy and look first for the common ground, the shared policy goals and objectives that might characterize the unity of environmentalists. To support this strategy, I will employ a useful, if somewhat arbitrarily drawn, distinction between *values* and *objectives*. An objective will be understood as some concrete goal such as a change in policy or the designation of a particular area as a wilderness preserve. Values will be understood more abstractly as the basis for an estimation of worth, which can serve as a justification and explanation for more concrete objectives. Thus two environmentalists might work together to achieve the objective of prohibiting strip mining in a wilderness area, while justifying their activities by appeal to quite different values. One of them might, for example, value the wilderness area as sacred, while the other wishes to perpetuate its recreational value for the use of the community. Differences in value may, therefore, lead to shifting coalitions regarding objectives; once strip mining is effectively prohibited, supporters of recreational values may find themselves allied with the local Chamber of Commerce in supporting a larger parking lot for access to the wilderness, while their former ally opposes both, insisting that ease of access will cheapen and degrade the sacred place.

Providing environmentalists can usually agree on what to do, a diversity of value concerns need not debilitate the movement. Indeed, freedom to appeal to a variety of value systems may ultimately prove the greatest strength of the movement, allowing environmentalists to appeal to the broadest spectrum of American voters.[12] Nevertheless, it is tempting to assume that one side or the other in the debate between Moralists and Aggregators is correct, and that there are some facts or theoretical arguments that will decisively vindicate one worldview or the other as expressing the correct vision to guide environmental policy. Most philosophers who have written on environmental ethics adopt this assumption, and have therefore debated the truth of "nonanthropocentrism," the view that nonhuman elements of nature have value independent of human values. Scientists, on the other hand, adopt the same assumption that we must choose between moralism and aggregationism, and expect that more scientific data and more sophisticated theories will determine what our approach should be.

The strategy of this book will be to think about environmentalism as a force in public policy first and to examine philosophical questions in passing, reflecting on questions of concepts, logic, and values and introducing ideas from philosophy only as they are necessary to understand policy debates, letting various ideal arguers—Aggregators, Moralists, and occasionally others—develop the best case they can for various policies and then examining their arguments. The book has three major sections. In Part I we will examine, in very broad historical terms, the first one hundred years of interactions of Aggregators and Moralists in the search for an environmental policy, a history that can be considered to have culminated in Earth Day 1970. Part II will examine current goals and beliefs of environmentalists in four broad policy areas: resource use, pollution control, protection of biodiversity, and land use policy. Based on the gradually developed philosophical ideas that have proved useful in discussing policy disputes in these areas, we will take a broad look at large-scale philosophical issues in the final part.

The most important consequence of this policy-oriented approach, and the rough distinction we have drawn between policies and the ideas that justify them, is that we will not miss examples where environmentalists pursue a policy by consensus in the policy arena, even while discussing and supporting these policies in quite different frameworks of concepts and values.[13] Another consequence of the approach is to hold open, throughout the inquiry, the possibility of a pluralistic integration of environmental values rather than an all-or-nothing decision between the Aggregators and the Moralists. It may be possible, given this approach, to escape the excruciating dilemma and construct an integrated approach to valuing the natural world.

Part One

The First 100 Years

2

Moralists and Aggregators:
The Case of
Muir and Pinchot

Gifford Pinchot first met John Muir in 1896, while on a trip through the West to study possible sites for new forest preserves. Pinchot was much impressed by Muir, twenty-seven years his senior, and recalled the meeting fifty years later in his autobiography. He described Muir as "cordial, and a most fascinating talker, I took to him at once."[1] Muir, in his writings of this period, was explicitly complimentary of Pinchot's efforts at sustainable forestry.[2] At the Grand Canyon, Muir and Pinchot struck off on their own and "spent an unforgettable day on the rim of the prodigious chasm, letting it soak in." They came across a tarantula and Muir wouldn't let Pinchot kill it: "He said it had as much right there as we did."[3]

Within a year, however, Muir had complained bitterly and publicly about Pinchot's decision to allow grazing in the national forest reserves.[4] This rift between the Moralist (Muir) and the Aggregator (Pinchot) shaped the two wings of the environmental movement, and its original configuration owes much to attitudes developed in the early life and work of each man.

The Early Odysseys of Muir and Pinchot

Muir entered the University of Wisconsin in 1861, the year the Civil War broke out. Although he was almost twenty-three, his last formal schooling had been interrupted at the age of eleven, when his family emigrated from Scotland. His father, Daniel, a religious zealot, had no use for any book but the Bible. The elder Muir, who joined ever more extreme sects in search of one sufficiently pure and exacting, chose eighty acres of virgin

17

land and put his eldest son John to work clearing it. Days were spent cutting trees and grubbing out roots, and nights were given over to memorizing Scripture.

Daniel Muir planted only corn and wheat for cash crops, and the farmland was worn out in only eight years. Choosing a new and larger plot, the family moved and repeated the process. Again, the hardest work fell to John as his father spent all of his time studying the Bible and preaching to anyone who would listen. Because of the topography of the new site, the only source of water was below seventy feet of rock. For months John was lowered into a well and spent each day chipping, with hammer and chisel, through rock. Once he was hauled up unconscious, almost dead from excess carbon dioxide in the damp hole. The next day he was back at his task and finally found water eighty feet below the surface.[5]

Young John's first break from farm drudgery came when, in 1860, he exhibited his inventions, mechanical devices carved in the middle of the night from scraps, at the Wisconsin State Fair in Madison. The inventions, including an "early rising machine" that tumbled its occupant out of bed and onto his feet at an appointed hour, were a sensation, written up in the newspapers as the highlight of the fair.

Once in Madison, understandably, he had little interest in returning to the farm, and enrolled at the university. Muir was fascinated with chemistry, physics, and especially botany. He enjoyed the mechanical process of dissecting plant specimens, applying the analytical tables, discovering the nature of the whole by analyzing the functions of its parts.[6] Botany became his new passion. Muir's knowledge, already spotty because he was largely self-taught, grew unevenly.

Still influenced by the religious fervor of his father, he shied from any contact with worldliness and rebuked his peers for irreverence and other sins. Considered an eccentric genius by his teachers and fellow classmates, he was accepted but held at a distance; he felt himself a social misfit.

Already alienated from society, Muir became paralyzed intellectually by obsessive worry that he would be drafted. With pacifist leanings and mainly Scottish allegiances, he wanted no part of what he considered a stupid, faraway war of someone else's doing; in 1863, facing almost certain conscription, he walked north through upper Michigan into Canada. He considered himself a fugitive and, avoiding settlements, wandered aimlessly in the trackless Canadian woods; his only solace was botany. He had brought along his plant press and his herbarium and spent his days collecting and preserving specimens. Although he later described these months as "botanizing in glorious freedom around the Great Lakes," he was also embarrassed and alienated by his flight.[7] Loneliness drove him into the arms of nature.

Then, in late June 1864, the vagabond had a new kind of religious experience, one that he later compared to meeting Emerson as one of the two high points of his life.[8] Wandering along the banks of a stream, he discovered two buds of the rare orchid, *Calypso borealis*. Seeing the plants there alone, as he was, in a huge wilderness, touched Muir deeply, "I felt as if I were in the presence of superior beings who loved me and beckoned me to come. I sat down beside them and wept for joy."[9] The serene beauty of the flowers, existing entirely independent of human manipulations or utility, caused him to question the human-centered assumptions of his Christian upbringing. At first the questions emerged in small, botanical ways: "I cannot understand the nature of the [Biblical] curse, 'Thorns and thistles shall it bring forth to thee.' Is our world indeed the worse for this 'thistly curse?' Are not all plants beautiful? or in some way useful? Would not the world suffer by the banishment of a single weed? The curse must be within ourselves."[10]

Gradually, from this experience forward, Muir began to see the world from a new, larger-than-human perspective. He eventually found a job as an inventor and machinist, rode out the war, and returned to the United States to accept another job in a machine shop. Then, in 1867, an accident with a file punctured his cornea, blinding his right eye. A few days later, sympathetic nerve blindness in the other eye left him sightless and desolate. Long, dark days and nights ran together as he despaired of ever seeing the beauties of nature again. Sight returned to his left eye and, much more slowly and against the predictions of the doctor, to the right eye as well. But he had lost interest in machine shops. Muir resolved to take a "grand sabbath day three years long," and explore the tropics. Taking a train to Louisville, he walked south toward Florida, with a somewhat vague plan to follow Humboldt's path into the Amazon in search of exotic plants.

Throughout this trip, the encounter with *Calypso borealis* worked on his mind. Sleeping in a Savannah graveyard, he questioned the Christian conception of death, which he found morbid: "Instead of the sympathy, the friendly union, of life and death so apparent in Nature, we are taught that death is an accident, a deplorable punishment for the oldest sin, the arch-enemy of life, etc. Town children, especially, are steeped in this death orthodoxy, for the natural beauties of death are seldom seen or taught in towns."[11]

He rejected monotheistic religion and the idea of human dominion as two aspects of the same flawed idea—Nature itself was the great creator; we, in our conceit, created a monotheistic God who cares only for us, a God fashioned in our own, conscious image: " 'The world, we are told, was made for man,' he noted. 'A presumption that is totally unsupported by facts.' "[12] "Nature's object in making animals and plants might possibly be first of all the happiness of each one of them, not the creation of all for the happiness of one. . . . The universe would be incomplete without man; but it would also be incomplete without the smallest transmicroscopic creature that dwells beyond our conceitful eyes and knowledge."[13] The new spiritual outlook therefore led also to a new understanding of individuality itself: "There is not a fragment in all nature, for every relative fragment of one thing is a full harmonious unit in itself."[14]

Muir's rejection of Christian anthropocentrism in favor of a pantheistic perspective, which he never included in his public writings,[15] emerged as a natural extension of his botanizing.[16] To Muir, the discovery of the orchids in Canada set in motion thought processes that upended his conception of the world. His painstaking collection of plant specimens, his discovery of the lonely orchids, and his acceptance of alligators as children of God were all scientific discoveries—biology and theology were but two manifestations of the same inquiry.

Muir's journey toward South America was interrupted by malaria; after a long period of rest he sailed to New York and left by boat, in steerage, through Panama, for California. Again, his goal was a botanizing trip to South America—he intended to stay in California for only a few months.[17] But a higher destiny was to intervene. Years earlier, a reputed "seer" had told John Muir that his destiny was to be found in the Sierra Nevadas, before Muir had heard of the mountain range.[18] Muir landed in San Francisco, got off the boat, and inquired of a passerby how he could get out of the city to a wild place. He was directed toward Berkeley, but he did not stop until he had arrived at Yosemite. There he became a shepherd and eventually ran a one-man sawmill. During his free time, he explored the mountains and its vegetation. He once lashed himself to the top of a tall pine during a violent thunderstorm in order to experience the storm more fully.

Muir met Emerson in 1871 and Emerson was much impressed by Muir's intensity and his prodigious conversational abilities. Muir failed, however, to persuade the sixty-eight-year-old Emerson to join him in a back-country camping trip. Muir's friends had for years urged him to write, but he found the task onerous. His first effort was to write down his observations and theories on the formation of the Sierra Nevada mountain range. This work was published in a leading New York newspaper; Muir became an instant, if reluctant, literary celebrity.

Then, in 1880, Muir married the daughter of a prosperous orchard owner, and for the next decade he successfully managed the family's farms and orchards. His writing and his wilderness rambling virtually ceased. Muir was eventually goaded out of this comfortable niche in 1889 by Robert Underwood Johnson, editor of *Century* magazine. Johnson extracted from Muir a promise to write two essays for *Century* in return for the magazine's support for making Yosemite a national park. Muir and Johnson succeeded not only in creating the second national park, but in quintupling its expanse.

Muir was launched in a second literary career. Although he wrote slowly and painfully, he found editors of all the East coast "polite" magazines anxious to publish anything he produced. Then, in 1892, Muir helped found the Sierra Club and served as its first president. "The battle we have fought," he wrote in the *Sierra Club Bulletin*, "and are still fighting for the forests is a part of the eternal conflict between right and wrong, and we cannot expect to see the end of it."[19] Muir had by then heard of vague plans for a dam in Hetch Hetchy valley, inside the confines of the new National Park. Already he was girding for an apocalyptic battle—his last—with the forces of evil.

If Muir was destined to become the leader of aesthetic and morally inspired nature preservation, the younger Pinchot—as different from Muir as a person could be—was trained and groomed to lead the forces of scientific resource conservation. Son of a wealthy and aristocratic family of French descent, Pinchot's father operated a successful dry goods firm in New York and was able to retire to a forested estate in Pennsylvania. In 1885 he asked his son, "How would you like to be a forester?" The younger Pinchot later recalled:

> As a boy it was my firm intention to be a naturalist. Camping was my delight. My pin-fire shotgun was my treasure. I had heard a panther scream in the Adirondack woods the summer my Father gave me my first rod and taught me to cast my first fly. . . . Of course a youngster with such a background would want to be a forester. Whatever Forestry might be, I was for it.[20]

Since there was no forestry profession in America, nor any formal course of study offered in the field here, young Pinchot entered Yale to study science: "Whoever turned his mind to forestry in those days thought little about the forest itself and more about its influences, and about its influence on rainfall first of all." Pinchot, in his autobiography, described his study at Yale: he took courses in meteorology, geology ("for forests grow out of the earth"), botany ("which has to do with the vegetable kingdom—trees are unquestionably vegetable"), and astronomy ("for it is the sun which makes trees grow"). He concluded: "All of which was as it should be, because science underlies the forester's knowledge of the woods. So far I was headed right."[21]

Upon graduation from Yale, Pinchot traveled to Europe, where the study of forestry was much more advanced. He met and fell under the tutelage of Sir Dietrich Brandis, a retired German forester of excellent reputation who was credited with having instituted forestry in India.[22] With Brandis's approval, he enrolled in the French Forest School at Nancy and plunged into his work with characteristic vigor. The assistant director of the Forestry School, Lucien Boppé, emphasized field experience, and greatly impressed Pinchot with his practical knowledge. Pinchot later said that he never lost sight of Professor Boppé's parting advice: "When you get home to America you must manage a forest and make it pay."[23] That advice became the motto of Pinchot's splendid career.

A formidable task faced him upon his return to America—his profession did not exist here. He summed up the American attitude toward forestry: "When the Gay Nineties began, the common word for our forests was 'inexhaustible.' To waste timber was a virtue and not a crime. There would always be plenty of timber and everything else in America for everybody, world without end."[24] Lumbermen, he said, "regarded forest devastation as normal and second growth as a delusion of fools, whom they cursed on the rare occasions when they happened to think of them. And as for sustained yield, no such idea had ever entered their heads."[25]

But Pinchot was a practical and remarkably able man; despite the obstacles, he succeeded in building and leading the fledgling forestry profession as a major force in America. Pinchot describes, in his autobiography, how he "discovered" utilitarianism more than a decade after becoming a government forester. Pinchot noted that, because much of Theodore Roosevelt's business with natural resource bureaus was conducted through him, "the Forest Service was dealing not only with trees but with public lands, mining, agriculture, irrigation, stream flow, soil erosion, fish, game, animal industry, and a host of other matters."[26] It was therefore natural, he noted, that the relations among these elements would occupy his mind. "But for a long time my mind stopped there. Then at last I woke up." Pinchot described how, on a particular afternoon in 1907, while riding his horse in Rock Creek Park outside Washington, D.C., he first recognized the connection among his diverse responsibilities. "Suddenly," he said, "the idea flashed through my head that there was a unity in this complication. . . . Here were no longer a lot of different, independent, and often antagonistic questions, each on its own separate little island, as we had been in the habit of thinking." He realized, that afternoon, that "here was one single question with many parts. Seen in this new light, all these separate questions fitted into and made up the one great central problem of the use of the earth for the good of man."[27]

Pinchot recognized that others before him had seen the importance of particular threats to the resource base, but thought he had here stumbled onto a grand unification of what had hitherto appeared as isolated problems: "So far as I know then or have since been able to find out, it had occurred to nobody, in this country or abroad, that here was one question

instead of many, one gigantic single problem that must
generations, as they came and went, were to live civili
lives in the lands which the Lord their God had give
discussed his new idea with his family and associate.
U.S. Geological Survey quickly grasped the idea and "becan.
brains of the new movement." It was McGee, according to Pinc..
"defined the new policy as the use of the natural resources for the great.
good of the greatest number for the longest time."[29]

Pinchot then discussed the idea with Roosevelt, who "adopted it without
the smallest hesitation. It was directly in line with everything he had been
thinking and doing, and became the heart of his administration."[30] After
long discussions with his associates concerning a label for the new idea,
one of them (Pinchot forgot which "and it doesn't matter") christened the
new policy "conservation." Roosevelt approved the label instantly.[31] The
term had been used as early as 1875 to describe a variety of activities ranging from forest preservation to saving water power.[32] Its choice as the label
of the Progressive movement's policy on natural resources represented little more than a consolidation of ideas already current, but the term "conservation" is now often used to refer to wise-use environmentalism to which
Pinchot gave his personal stamp. There is no question that Pinchot and
McGee gave the policies a distinctively utilitarian formulation and explanation. Roosevelt, who had originally been as close to Muir in his ideas as
to Pinchot, was carried along in this current among his advisers.[33]

Muir, by chance or destiny, fell into the role of protector of the Sierra
landscapes. Pinchot, by following a well-orchestrated plan, was to become
the most powerful voice speaking for the scientific development of natural
resources. The working out of these destinies flowed, in part, from the role
each was to play in the development of the turn-of-the-century American
"land-saving" organizations.

Early Activism

Concern for environmental degradation in America had been implicit in
the writings of Henry David Thoreau and the other transcendentalists, but
their intense individualism largely precluded activism. The publication, in
1864, of George Perkins Marsh's classic *Man and Nature* (though it was
initially unnoticed because of the Civil War) marked a different approach.
Marsh, in a systematic analysis, exposed the wasteful aspects of the dominant trend of development and conversion of wilderness into civilized, productive land. A curious mixture of acute scientific observation and religious
speculation, Marsh's remarkable book was representative of scientific thought
in the era immediately preceding Darwin's revolutionary influence. Sci-

~e had largely demystified the physical world, replacing demons with ,quations and mystery with mechanism. But life, consciousness, and order in the biological realm had not yet been brought within the scientific worldview by full acceptance of Darwinism.

Two quite independent forces developed in response to these trends. One group, which included Charles Sprague Sargent, director of the Arnold Arboretum at Harvard, and Robert Underwood Johnson, editor, as mentioned earlier, of the influential *Century* magazine, eventually included Muir, who was introduced to Sargent in 1878.[34] They sometimes called themselves "Friends of the Forests," and generally favored setting aside forest preserves in public ownership, prohibiting all cutting and economic activities in them, and supported the use of army personnel to enforce this protection. This group was able to build a broader constituency by appealing to an awakening interest, especially among well-educated and patrician Easterners, in nature and the out-of-doors.

Truly activist conservationism was initiated by the Alpine clubs, such as the Williamstown (Massachusetts) Alpine club founded in 1863, and by ornithological and sportsmen's organizations founded in subsequent decades.[35] These developments reflected a romantic reaction on the part of affluent urbanites, who began to lament the loss of wilderness and to develop a romanticized aesthetic sense of its value. Members of the Alpine clubs, the prestigious Boone and Crockett Club, and other sportsmen's clubs adopted nature protection as a hobby. As Stephen Fox described them, they were members of "the anxious gentry [who] hoped to recover the lost manly virtues—courage, self-reliance, physical strength and dexterity—and so avoid the spectre of 'race suicide.' "[36] This fad for nature and wilderness was a distinctly upper-class phenomenon, accompanied by a wave of nature essays and adventure chronicles in the "polite" magazines of the day. It was these groups and organizations that provided the constituency and the force for the campaigns of Sargent, Muir, and their followers.

Believing that scientific and economically responsible management would be the wave of the future in conservation, Pinchot decided not to join with the Friends of the Forests. He therefore mounted a second force by building forestry as a profession. The Friends of the Forests, he thought, had adopted a losing strategy: "They tried to stop the advance of one of the greatest, most necessary, and most thriving and driving of industries simply by explaining to each other how wrong and ruthless it was." Pinchot, who had been taught in Europe that to be practicable forestry must yield economic rewards, concluded that moral preachments would be ineffective. He believed that Sargent and his amateur followers would fail because "their eyes were closed to the economic motive behind true Forestry."[37] He noted that they were ridiculed as "denudatics," and said "the lumber juggernaut rolled over them—rolled over them and went on its forest-devastating and homebuilding way without even paying them the tribute of serious attention."[38] Pinchot concluded that, in spite of laudable goals of forest preservers, not a single acre of publicly or privately owned forests were systematically managed.[39] Of Muir and his cohorts, he said: "They hated to see a tree cut down. So do I, and the chances are that you do too. But you cannot practice forestry without it."[40]

Pinchot's attitude about what counts as "management" is illustrated by his remarks on the Adirondacks: "In 1885 New York State created the Adirondack State Forest Preserve, not for Forestry, but to 'preserve' the forest and the water supply."[41] In short, the effort to " 'preserve' the forest and water supply" was not counted, by Pinchot, as management at all. Because he had identified his personal interests in building a forest service with a particular style of economically motivated forest management characteristic of European forestry, Pinchot dismissed the New York effort as not really qualifying as management at all.

The utilitarian criterion instructs us to maximize three variables simultaneously—wealth, democratic distribution of wealth, and sustainability of resource use over indefinite time. Because it is impossible to maximize more than one variable at once, applications of the principle can take a number of different forms, depending on which variable is emphasized at a given point. This feature of the utilitarian criterion provided Pinchot with a "solution" to his political dilemma of how to convince citizens, especially citizens of Western boom states, that conservation was in their interests. He emphasized, in political contexts, the second variable; he argued that conservation meant democratic distribution of the fruits of resource development. He was therefore able to sell conservation as a means to halt the destructive practices of the timber barons and other wealthy private exploiters. He indiscriminately attacked concentration of wealth and wastefulness in production as if they were two heads of the same monster, but his intellectual solution allowed him political slack; he could choose to emphasize the variable that was politically expedient in a given situation.

Pinchot, in effect, equated happiness, in the utilitarian formula, with material well-being. Timber production and grazing, therefore, were considered legitimate "uses" of the national forests, because they had an immediate payoff in the production of material goods. Combining the utilitarian formula as a criterion of good conservation practice, European training in production forestry, and a powerful political need to emphasize growth and development, Pinchot developed a narrow viewpoint of the public good

he served—a viewpoint that greatly overemphasized short-term development of material resources at the expense of longer-term and broader values such as aesthetics.

Despite their different orientations, the Friends of the Forests and Pinchot at first worked together. Under the influence of Marsh's important
book, they began pressing for scientific study of the forest situation, and
they agreed that retention of large tracts of forest land in public ownership
was extremely desirable. Acting on these shared goals, they worked to improve the climate for forest protection in America. The American Association for the Advancement of Science, in 1873, encouraged President Grant
to hire one person in the Department of Agriculture for the purpose of
gathering statistics and distributing information on forest use. By 1881, a
tiny Division of Forestry was in existence in the Department of Agriculture, but its purpose was informational only; actual forest management,
while often discussed, was virtually nonexistent.[42]

Given his pessimistic assessment of the climate for professional forestry
upon his return from France, Pinchot concluded he would have to blaze
his own trail.[43] Acting on this instinct, which was reinforced by a personality conflict with the chief of the fledgling Forestry Division, Pinchot rejected a position as assistant chief and accepted the challenge of experi-
mentally managing 5000 acres of land on the Biltmore estate of George W.
Vanderbilt. Frederick Law Olmsted, who was employed by Vanderbilt to
develop his Asheville, North Carolina estate into the greatest country estate in the nation, suggested the experiment.[44] This offer provided Pinchot
with an opportunity to follow the advice of his European mentors, Boppe
and Brandis: to prove that forest management was economically feasible.

As he worked, with moderate success, to make the Biltmore forest pay,
he moved also to improve the public climate for forestry. Working with
Johnson, Pinchot built congressional support to establish a forest commission. In 1896 the Department of the Interior requested that the National
Academy of Sciences establish a study group. The National Commission,
to be led by Sargent with Pinchot as secretary, was established that same
year.

Pinchot was the only professionally trained forester on the commission
(the other members were academically oriented scientists) and he soon
clashed with Sargent on three major points.[45] First, Sargent moved slowly,
despite the fact that President Cleveland and his Secretary of the Interior,
Hoke Smith, were anxious to put recommendations before Congress during
the 1896 session. The impatient Pinchot chafed under the deliberate methods of Sargent. Second, Cleveland wanted, most of all, a plan for running
forest reserves—a management plan for a small system of reserves that
could be expanded later. Pinchot was highly critical of Sargent for moving

first to recommend a major expansion of reserve areas without including a plan for managing either them or existing reserves. Third, Pinchot and Sargent differed over Sargent's plan to use army personnel, trained at West Point, rather than proceeding immediately to develop a strong and independent corps of trained foresters.[46]

As the end of Cleveland's term approached, the Commission's report was not completed, although the members had traveled throughout the West inventorying sites for possible reserves. In order to provide the President with some basis for action, an interim report was forwarded to Smith recommending that the current forest reserves be supplemented with thirteen new ones that would more than double the current acreage in reserves.[47] When Cleveland acted to set aside the new reserves by presidential order, negative reaction from the West was immediate and violent. Sargent, who once confided to Muir, "We have got to act promptly and secretly in these matters, . . . or the politicians will overwhelm us,"[48] preferred to work in private, so the Commission's work had not been publicized; no attempt had been made to explain or gain support for expanding the reserves.

Pinchot saw the problem in political terms:

> The creation of thirteen new Reserves in seven States came like a thunderclap. And since under existing interpretations of law no use whatever could be made of the resources of the old Reserves, or of the new, since even to set foot on them was illegal, the only possible conclusion was that this vast area was to be locked up, settlers were to be kept out, and all development permanently prevented. No wonder the West rose up.[49]

Demonstractions occurred throughout the West. Settlers in the Black Hills of South Dakota dressed up in war paint and staged mock attacks on trains to protest what they saw as a plan to retard the development of the West and return it to the Indians.[50] In the subsequent conflagration, Pinchot gained further education in pragmatic politics and, although congressional action reopened for use all but two of the new reserves, legislation signed by the newly inaugurated President McKinley included provisions for managing current and future reserves.

As the work of the Forest Commission continued, Pinchot and Sargent were increasingly at odds. Sargent and most of the Commission wanted to keep the reserves inviolate and to use the army, not professional forest managers, to protect them.[51] Pinchot argued valiantly for economically useful reserves under the direction of forest managers. In the end, Sargent's views dominated the report and, after considering filing a minority report, Pinchot reluctantly signed with the majority in return for small concessions toward the future goal of professional forest management.[52]

Sargent's worst fears—that his rival would enact his minority opinions at the expense of the larger Commission's report—came true when Pinchot

accepted a position as "Special Forest Agent" for the Department of the
Interior. After completing a report arguing for employing professionally
trained forest officers, Pinchot was hired as the new head of the Forestry
Division in the Department of Agriculture. Although the forest reserves
were under the jurisdiction of the Interior Department, Pinchot's recommendations affected their management and eventually, with the help of
Roosevelt, Pinchot engineered the transfer of responsibility for the reserves to the Department of Agriculture.

Pinchot proceeded, under McKinley and with increasing success under
Roosevelt after McKinley's assassination, to develop a strong, active, and
professional Forest Service. This goal was furthered when his wealthy family made a large gift to Yale University to begin the nation's first school of
forestry. His objectives, while pursued against a backdrop of moral concern
for resource use, were thoroughly pragmatic. After McKinley's death and
even before Pinchot's friend Roosevelt could move into the White House,
Pinchot took F. H. Newell to visit the incoming President. Newell, an
engineer in the Geological Survey, was an advocate of irrigation and he
and Pinchot were invited to submit recommendations for Roosevelt's legislative initiatives. In just six months, the Bureau of Reclamation had been
instituted and Pinchot's influence began to spread beyond forestry and into
the other bureaus of the federal government.[53] The transfer of the Forest
Reserves, Pinchot's personal goal, took a little longer. But in 1905, Congress authorized the transfer of the reserves to Agriculture and their name
was changed to "National Forests." They were placed under Pinchot's Bureau of Forestry, which was renamed the United States Forest Service.
The Secretary of Agriculture immediately sent a letter (drafted by Pinchot
himself) to Pinchot directing that "all land is to be devoted to its most
productive use for the permanent good of the whole people, and not the
temporary benefit of individuals or companies."[54] Pinchot, apparently, could
not conceive any possible conflict between the goals of using land for its
"most productive use" and protecting "the permanent good" of the people.

It is impossible to overestimate the importance of Pinchot's subsequent
influence on forest policy and on the development of other conservation-
oriented agencies as part of Roosevelt's progressive program. For example,
Pinchot was authorized to write and present a message to an irrigation
congress in Roosevelt's name, with assurance from the President that "he
would stand for whatever [Pinchot] said."[55]

Sargent and Muir, with some justification, felt they had been sold out.[56]
It especially rankled them that they had assisted and advised the ambitious
Pinchot in his career. Aggravated by these personal and professional clashes,
Sargent and Muir battled Pinchot over his relentless attempts to find

immediate economic uses for the forest preserves, especially when these
economic uses came at the expense of aesthetic considerations and other
less obviously commercial goals such as watershed protection.

Both Pinchot and the Friends of the Forests favored "reserving" large
areas of forest in public ownership, clashing only regarding their management. Pinchot's general attitude is illuminated by the account in his autobiography, entitled "Not All Sheep Are Black," of the grazing issue: "The
most ticklish question Secretary [of the Interior Ethan] Hitchcock had asked
our advice about in 1899 was grazing. I was already in touch with the
Western sheepmen and irrigators, and I knew it would be a tough nut to
crack." He recognized that "grazing was the bloody angle, and obviously
was to be for years. It was the most important use that had yet been made
of the Forest Reserves, and the center of the bitterest controversy." Referring immodestly to a 1900 fact-finding trip, he "put in three weeks that
made history for the grazing industry of the West."[57] "I kept . . . [my]
eyes open wherever we went, with the sheep question uppermost," he
said. "This trip established what I was sure of already, that overgrazing by
sheep does destroy the forest. Not only do sheep eat young seedlings, as I
proved to my full satisfaction by finding plenty of them bitten off, contrary
to the sheepmen's contention, but their innumerable hoofs also break and
trample seedlings into the ground." He even cited Muir on sheep: "John
Muir called them hoofed locusts, and he was right."[58]

But he did not conclude, with Muir, that grazing should be stopped in
the Forest Reserves: "Great stretches of open forest contain much feed that
should not be wasted, provided the ranges are not overstocked and provided again (and this is of the first importance) that when reproduction of
the forest is needed, grazing stops. . . . Exclusion may be necessary for
from about a tenth to a fifth of the time it takes to grow a merchantable
tree." And thus Pinchot opted for enforcement of rules, not exclusion of
grazing: "Sheep grazing in the forest requires special care. Every sheep
owner should have the sole right to his range for a reasonable time and at
a reasonable charge, but overgrazing should forfeit his permit and the money
he paid for it."[59]

His report to Secretary Hitchcock was clear: "To regulate grazing is usually far better than to forbid it altogether." He later justified his recom-
mendation:

> In the early days of the grazing trouble, when the protection of the public
> timberlands was a live political issue, we were faced with this simple choice:
> Shut out all grazing and lose the Forest Reserves, or let stock in under
> control and save the Reserves for the Nation. It seemed to me there was
> but one thing to do. We did it, and because we did it some 175,000,000

acres of National Forests today safeguard the headwaters of most Western rivers, and some Eastern rivers as well.[60]

While hardly modest, these are the explanations of a practical, not doctrinaire, man. Pinchot did what, in his opinion, had to be done to save forests for forestry, to promote a strong forestry profession, and to protect the Forest Reserves in public ownership. He essentially traded development rights (timber, mining claims, grazing rights, and so on) to buy off various constituencies in the western states and territories. If grazing sheep would help to make the forests profitable (and enhance critical support for the forests in the West), then he was inclined to find some means to make that use compatible with good forest management.

The final break between Pinchot and Muir occurred over the plan to dam Hetch Hetchy Canyon in Yosemite National Park, which deeply split the early forces for environmental protection. Some Sierra Club members supported the dam, including Warren Olney, one of its most influential founders. When the club took a vote to determine its position on the dam, 161 members supported the dam and 589 opposed it. John Burroughs, Muir's close friend and fellow nature writer, felt that Muir was just being crotchety about the dam, saying: "Grand scenery is going to waste in the Sierras—let's utilize some of it."[61]

The historian Stephen Fox concludes: "Granted that sincere conservationists could disagree about the project, the real division . . . was occupational, between those who urged the dam *as part of their jobs* and those who *took time from their jobs* to oppose it. In short, another collision of professionals and amateurs."[62] The main proponents were employees, elected and appointed, of the City of San Francisco. Residents, partly in retaliation against the inefficient and hated Spring Water Company, voted 7 to 1 in favor of the dam. The employees of the city therefore felt they had an overwhelming mandate to pursue the project.

Muir and the amateurs were at a serious disadvantage. While Muir was by this time secure enough financially to devote time to the struggle, his younger colleagues were forced one by one to withdraw or greatly curtail their activities. The battle, however, was not one-sided. Most articulate public opinion nationwide, excluding residents of San Francisco, opposed the dam on aesthetic grounds.

At one point, when Taft succeeded Roosevelt and visited Yosemite with Muir, the preservationists thought they would prevail. Taft and Secretary of Interior Ballinger were involved in a rancorous dispute with Pinchot, which eventually cut short Pinchot's career in the Forest Service. Since Pinchot was identified with the dam, they were inclined to oppose it. Delays ensued and Muir used the time to build a national network of preservation-oriented followers who opposed all development in the National Parks. In the end, however, politics decided the issue. After Wilson's election in 1912, he appointed Franklin Lane, a city attorney active in San Francisco's fight for the dam, as his Secretary of Interior, apparently as a political reward for Lane's effective campaign to carry San Francisco for Wilson. With the backing of Wilson and Lane, the proposed dam received congressional approval in 1913.

Defeated and discouraged, Muir weakened, suffered several illnesses, and died a year later. His influence continued because of the powerful network of amateurs that he and his friends had created in the attempt to stop the dam. This group could be counted on to support preservation issues, and led the battle for game protection. Pinchot, who had been discharged from the Forest Service in 1910 as a result of the bitter controversy with Taft and Ballinger, continued to have a strong influence because of the many foresters and resource managers he had placed in important posts throughout the government and also because he led the National Conservation Association until 1922, when he drifted away from direct work in conservation to run for political office. He later became governor of Pennsylvania.

At the end of the first decade of this century, the movement to "save" nature was deeply split by the Hetch Hetchy controversy. Pinchot's followers were in ascendance; they occupied powerful positions in the newly developed and expanding federal bureaus intended to "manage" the nation's resources. But the Sierra Club and other private organizations represented a sleeping political giant, ready to be mobilized from time to time in national crusades. Although the opponents of the dam had lost their battle to save Hetch Hetchy, they had created a powerful legacy by developing a national constituency and by learning the techniques of political influence. That influence was felt when Wilson and Lane, perhaps in order to balance the Hetch Hetchy verdict, acted on a long-standing request of the Sierra Club for a unified agency, similar to the Forest Service, to have responsibility for the National Parks. Stephen Mather, a Sierra Club member and opponent of the dam, was chosen to head the new National Park Service.

Ecstatic Science vs. Scientific Management

Most commentators who have analyzed the disputes between Muir and Pinchot have characterized them as representing a clash of conflicting value systems—utilitarian conservationists versus the nature-worshiping preservationists—and there is no doubt that Muir and Pinchot emphasized different goals and expressed polarized values in their public rhetoric. In this sense, Muir and Pinchot provided prototypes for the Moralists and Aggregators who personify the polarization in values that dominates the landscape among today's environmentalists.

The philosophical oppositions between Muir's and Pinchot's followers were by no means as simple and clear-cut as the rhetoric implied, however. The Sierra Club did not oppose a generally human-oriented utilitarianism, but often argued against short-sighted and overly materialistic interpretations of the public good. For example, Joseph LeConte, with Muir one of the cofounders of the Sierra Club, expressed this approach in an early issue of the Sierra Club *Bulletin:* "It is true that the trees are for human use. But there are aesthetic uses as well as commercial uses—uses for the spiritual wealth of all, as well as the material wealth of some."[63]

Pinchot, on the other hand, also expressed aesthetic, nonmaterial values. An avid hunter and fisherman who sincerely loved the outdoors, he was proud of his long and taxing camping trips. When Pinchot encountered "the gigantic and gigantically wasteful lumbering of the great Sequoias, many of whose trunks were so huge they had to be blown apart before they could be handled," he said, "I resented then, and I still resent, the practice of making vine stakes hardly bigger than walking sticks out of these greatest of all living things."[64] So Pinchot was not hostile to the aesthetic and moral concerns of Muir and his friends—Pinchot's conception of "scientific" resource management and his training in production-oriented forestry narrowed his focus and caused him to emphasize the contributions of forests and forestry to human *material* well-being.

Muir, on the other hand, never adopted the blinders of any particular profession. When asked, Muir referred to himself as a scientist.[65] To understand the clashes between Muir and Pinchot, I would argue, it is more important to understand their differing concepts of *science* than their differing concepts of *value*. Muir was a scientist not in the rising tradition of scientific resource management, but in the older tradition of naturalists, traceable to John Bartram (who was appointed King's Botanist for North America in 1765). The geographer Marsh, only thirty-seven years older than Muir, provided an excellent example of this less specialized scientific tradition. Eugene Hargrove has recently illuminated this tradition by tracing the productive interactions between field naturalists and nature artists in nineteenth-century America and by showing the extent to which artists and naturalists cooperated to create an ethic based upon combined scientific and aesthetic attitudes. This tradition, which valued species and landscapes (as links in the great chain of God's creation) rather than individual specimens,[66] emphasized the aesthetic *and* scientific interest of nature.

John Muir's work must be understood as a further development in this naturalistic tradition, which came to believe that everything natural has its own beauty, and which saw an organicist/pantheist theology as providing metaphysically "deeper" explanations for scientific facts. Aesthetics, science, ethics, and religion formed a unified system of knowledge. Muir extended the idea of Emerson, that "there is no object so foul that intense light will not make [it] beautiful," by applying it to parks and forest preserves. Muir argued that "none of Nature's landscapes are ugly so long as they are wild."[67] Another important contribution to the new aesthetic was the development in geology, beginning with Hutton and Lyell and culminating in the thought of Darwin, of the idea that biological processes occurred very gradually and over immense periods of time.[68] The idea that the gradually developed features of the natural world are sublime in some profound sense, and that scientific "interest" was as aesthetically important as beauty, had the effect of further melding scientific and aesthetic thought and of giving an unquestioned value to things simply because they exist.[69]

Building on Hargrove's insightful historical account, we can therefore begin to understand Muir's characterization of himself as, first and foremost, a *scientist*. Muir's pantheistic philosophy informed his conception of science: To learn a new scientific fact was to see, more clearly, the face of God, and to see the face of God was to discover yet more facts. Muir's conception of the great, beautiful, and interrelated world of man, nature, and creator was seamless; there was no separation of science from theology, of hard facts from aesthetic rapture, of facts from moral insights. Muir the wandering scientist and Muir the theologian were engaged in a single, ecstatic enterprise.

In historical context, Muir's claim to be first and foremost a scientist was not unreasonable. Muir is credited with the considerable geological discovery that the most visible features of the Sierra Nevada Range resulted from glacial activity; he also participated in the verification of his theory by discovering an active glacier, in Alaska, later to be called "Muir Glacier," in the process of gouging out a hitherto undiscovered bay.[70] While he used a vocabulary drawn from theology rather than science, some of his observations on natural systems anticipate ideas that were to become current in ecology decades later. For example: "The antipathies existing in the Lord's great animal family must be wisely planned, like balanced repulsion and attraction in the mineral kingdom."[71] Later, he was to write: "Every atom in creation may be said to be acquainted with and married to every other, but with universal union there is a division sufficient in degree for the purposes of the most intense individuality; no matter, therefore, what may be the note which any creature forms in the song of existence, it is made first for itself, then more and more remotely for all the world and worlds."[72]

Muir accepted the scientific aspects of Darwin's theory, but refused to call the process a "struggle," rejecting the word as "ungodly." He insisted, "Somewhere, before evolution was, was an Intelligence that laid out the plan, and evolution is the process, not the origin, of the harmony."[73]

Applying today's standards, and even the developing standards of scientific objectivity of his own day, it would be easy to dismiss Muir's claim to be a scientist, arguing that he was disqualified by his religious rhetoric and by his unorthodox methods. A scientist friend of Muir's, for example, said, "To me Muir was a poet rather than a scientific man; . . . he loved Nature and spent his life observing her works; but his observations were the observations of a poet, of a lover, and not the systematic observations of a man of science." Josiah Whitney, the state geologist of California, was less kind. A proponent of the dominant theory of the day, that the Sierras were thrust up by a cataclysm such as an earthquake, Whitney dismissed Muir as a "shepherd."

The amateur Muir, though he missed some details, proved much closer to correct than his professional rivals, however. This galled them the more because he had no use for their technical jargon and complex measurements, and patronized their attempts at "objectivity" as self-defeating: "This was my 'method of study,' " he told a meeting of geologists, "I drifted about from rock to rock, from stream to stream, from grove to grove. Where night found me, there I camped. When I discovered a new plant, I sat down beside it for a minute or a day, to make its acquaintance and hear what it had to tell. . . . I asked the boulders I met from whence they came and whither they were going."[74] Muir rejected the dominant theory because he believed the Sierras could not have been an "accident," concluding that nothing so beautiful could have been caused by a random cataclysm. Muir sensed the youth of the mountains and described them as "landscapes fresh from God's hand. . . . As if this were the dawn of creation and we had been blessed in being made spectators of it all." This metaphorical line of reasoning led Muir to identify glaciers as God's sculpting instruments.

Given the direction of science even in his day, Muir represented, by his willingness to mix religion, aesthetics, and science, a dying tradition in science. It was a rich tradition that had culminated in Marsh's pioneer classic on human abuse of land, and Muir had a legitimate claim to be called a scientist within that unspecialized, naturalistic tradition. But the tradition was swept away, even in his lifetime, by a new reductionistic and value-free approach to science.

Muir's aesthetic, moral, and theological ideas united in a scientific naturalism undergirded by his pantheism. Muir conceived an ecstatic science, a science motivated by a religious desire to know the face of God. Values and facts were thus inseparable. To know nature was, for Muir, to love nature. Science and theology were two sides of the same ecstatic coin. Muir bequeathed to preservationists a deep respect for holistic explanations and a longing for a broader science that did not require the banishment of value and religious awe from the observational process. He also followed Thoreau in emphasizing the power of observation of nature to transform a human individual and to convert that person from materialistic consumerism to joyous participation in the creativity and sustenance of life.

But Muir never denied that nature could and should be used to fulfill human utilitarian needs. A talented machinist and a successful orchard manager, Muir clearly recognized the importance of producing goods for human use. He realized that his attempts to preserve the most beautiful of God's landscapes depended on protecting the productive potential of lands already under use.

After President Cleveland had increased the Forest Reserves in 1897, the editor of the *Atlantic Monthly* asked Sargent to recommend a writer to defend the Forest Reserves. Sargent said, "There is but one man in the United States who can do it justice, and his name is John Muir!" Muir's article said the pioneers had treated "God's trees as only a large kind of pernicious weeds," and waged "interminable forest wars." Muir, the scientific mystic, advocated "rejoicing in wildness,"[75] But he also praised the concept of sustained yield, explicitly cited Pinchot's approach to "wise management," and took European forestry as a model, insisting that optimal conditions exist when "the state woodlands are not allowed to lie idle [but] . . . are made to produce as much timber as is possible without spoiling them." He concluded that careful selection of mature trees for cutting would ensure that the forests would remain "a never failing fountain of wealth and beauty."[76] When defending this line, Muir's approach differed from Pinchot's only in emphasizing beauty as well as wealth, and in seeing the forests as a spiritual resource for people. As their positions became polarized during the Hetch Hetchy battle, however, Muir mentioned wise use less and less. In opposition to the materialists, he gradually adopted a more consistently spiritual rhetoric.

Muir, in sum, propounded a rich, if unsystematic, worldview, one embracing a variety of human-utilitarian, spiritual, and human-independent moral values in nature. He saw these values as unified by his pantheism, which allowed him to tailor his rhetoric to the task of enlisting citizens in the goal of preservation. But that same imprecision doomed early preservationists to disrespect within the scientific community, which generally tried to isolate its descriptive work from moral and theological musings.

Pinchot, on the other side, had a strong, if somewhat intellectually and professionally underdeveloped, sense of natural aesthetics. The operative difference between the two great leaders of early conservation, therefore, was not simply adherence to contradictory value positions. Guided by differing emphases on a range of shared values, Pinchot thought science should be value-free while Muir scoffed at the idea. For Muir, science was a grand enterprise of understanding Man, Nature, and God—utilitarian applications of scientific knowledge were important, but important in a very narrow, practical sphere. For Pinchot, the pragmatic politician, goals of productivity were dominant, and he saw his professional task as one of developing resources—and he thereby assigned his broader aesthetic understanding of nature a relatively unimportant, nonprofessional role in his thinking. The competing value systems of Muir and Pinchot, the precursors of today's diverging worldviews of environmentalists, represent, then, not so much *incompatible* systems of valuing, but systems emphasizing different portions of a shared spectrum of values.

If Muir and Pinchot are portrayed as sharing a spectrum of values that they emphasized differently in different situations, rather than pursuing inconsistent values that are always in conflict, it is much easier to understand their considerable cooperation. This approach also shifts attention away from the philosophical question of which value system is more defensible in the abstract, to a series of questions about why Pinchot and Muir emphasized different values in different situations. Why, for example, was Muir adamantly opposed to mining and building reservoirs in the national parks, whereas Pinchot thought it appropriate to pursue these useful activities as long as steps were taken to minimize impacts on recreational uses? Why did Pinchot support regulated grazing in the forest reserves, while Muir thought this use entirely inappropriate?

A Partioned World?

It is not entirely fanciful to suggest that, had it not been for their personal differences, Pinchot and Muir might have worked out a more or less explicit pact—Pinchot would be the Minister of Wise Use while Muir would be the Minister of Aesthetic Appreciation. Pinchot would manage the National Forests and Muir would work to protect National Parks against degradation from economic activities. Because Muir never believed that *all* natural areas should be preserved, recognizing that some would have to be managed for human use, his worldview never explicitly conflicted with Pinchot's. If two contrary-to-fact conditions were fulfilled, Muir and Pinchot might, with no real change in their worldviews, have worked together in harmony. In a perfect world in which land resources were unlimited, there being no competition between economic and aesthetic uses and in which productive uses did not have spillover effects on nearby pristine lands, Muir and Pinchot might have divided the landscape into private lands, government-owned productive reserves, and government-owned aesthetic and recreational preserves. Pinchot would manage the productive preserves and Muir would protect the aesthetic preserves. There would be, on this partitioning scheme, no direct conflict between Muir's preservationist worldview and Pinchot's conservationist one. Indeed, some such partitioning was recognized politically when the National Park Service was set up, as a balance to the Forest Service in the Department of Agriculture, in the Department of the Interior.

In the actual world of political action, however, such a partitioning could never occur. Prime timber and minerals existed on or near aesthetic landmarks, and conflicts arose regarding which, and how many, areas would be designated as parks and preserves. The battle over Hetch Hetchy was precipitated because wise-use conservationists, including Pinchot, attempted to extend their wise-use strategy into Yosemite National Park. Similarly, the controversy over grazing was a question of how much grazing should be allowed in the National Forest Reserves. Pinchot's blind spot for aesthetic values played an important role in creating conflicts, because he seemed unable to see the importance of preservation for aesthetic purposes, but the real source of conflicts arose from the closing of the frontier. Once it became clear that there was a limited supply of land for all uses, conflicts arose regarding the appropriate use of particular parcels.

It could be said that the differing worldviews of Pinchot and Muir caused these disputes, but it would be more accurate to reverse the causal explanation—Muir and Pinchot, in the pursuit of their divergent policy goals came to emphasize quite different values. Pinchot, especially, ignored the values of aesthetic preservation that motivated Muir and his followers; but Muir, also, increasingly avoided reference to wise use. This bifurcation occurred because they disagreed about particular uses of particular parcels of land and found it useful to appeal to different constituencies with quite different values. Conservationists, who emphasized the utilitarian values of wise use, attempted to extend their style of management almost everywhere. Preservationists such as Muir, who never denied the importance of wise use in areas devoted to human productivity, emphasized that human values are broader than material ones, and pressed aesthetic and moral arguments to limit the seemingly inexorable expansion of management for economic productivity.

But Muir's concern to protect wild places of natural beauty shared one thing with Pinchot's reaction to waste of resources: that both were essentially *reactive*. Muir did not first develop a coherent philosophy and then deduce a preservationist policy from it. He rather set out to save landmarks from the advancing forces of greed and developed a complex ideology to justify his preservationist policies. His worldview encompassed Pinchot's utilitarian values, but balanced them with powerful aesthetic and moral commitments. What was lacking in Muir's worldview was a principle for integrating these richly diverse values. Thus, while Muir introduced an alternative worldview to compete with those of the exploitationists and Pinchot's conservationists, the positive values of the preservationist viewpoint were not defined in a philosophically unified way.

Both Muir and Pinchot left powerful legacies. Both created a distinct countercurrent against the exploit-and-desert tactics of early approaches to resource development. Both developed a general way of looking at the world, tentative steps toward a worldview, and a policy of protection. Muir emphasized the preservation of pristine areas; Pinchot advocated wise use of areas already under human use. Their disagreements often originated in questions about whose values and whose management plan was appropriate for a given area, and often led to rancorous disputes. But in spite of these disagreements, they co-created a movement: Because each of them, and their followers, effectively voiced a particular set of values, they provided overlapping, even politically complementary, options to exploitationism.

3

Aldo Leopold and the Search for an Integrated Theory of Environmental Management

The Young Forester's Dilemma

Aldo Leopold led two lives. He was, in the best tradition of Gifford Pinchot, a forester and a coldly analytic scientific resource manager, devoted to maximizing resource productivity. But Leopold was also a romantic, who joined the Forest Service because of his love for the outdoors, a love he never lost or fully subjugated to the economic "ciphers" that so constrain public conservation work.

During the last decade of his life, Leopold the romantic fashioned a little book of essays. He chose from the best of his stacks of field journals and his voluminous publications a few short essays, supplemented these with new pieces, polished them, and strung and restrung them like pearls. The manuscript, representing the essence of his long career, was given final acceptance by Oxford University Press only seven days before Leopold's death, and the essays were published as *A Sand County Almanac*. The final essay in that book is "The Land Ethic," which, Leopold said, "sets forth, in more logical terms, some ideas whereby we dissenters rationalize our dissent."[1]

Although he was not primarily an abstract thinker, Leopold, I will assert, has been the most important figure in the history of both environmental management and environmental ethics. This evaluation is based on one reason: Having faced the environmentalists' dilemma and, having to formulate goals and actions, he articulated a workable, practical philosophy that transcends the dilemma. The story of how he did so is a sketch of his life.

Leopold was a forester in the Southwest for fifteen years. He *saw* the range deteriorate. He *saw* the main street of Carson City erode into a deep chasm, and he knew, by the early 1920s, that his agency and its Pinchotist philosophy was significantly responsible. But he was as befuddled as anyone else, and grasped at philosophical straws, or any other straws, to articulate in general terms what was going wrong.

Leopold had entered the Forest Service at the height of the Hetch Hetchy controversy. He recognized, of course, that there were critics of the service, and he surely had some respect for Muir's viewpoint. But Leopold was a well-trained forester; while innovative and sometimes out of step with his colleagues, he generally went by the book.[2]

In 1923 Leopold penned a remarkable little essay called "Some Fundamentals of Conservation in the Southwest" (which was not published until 1979 in the first volume of the journal *Environment Ethics*). In that essay, Leopold began by recounting his experiences of watching the land systems of the Southwest failing before his eyes, then offered and reflected upon theories about why it was happening, and concluded with a broad-brush sketch of the philosophical alternatives to Pinchotism.

In his reflections, Leopold evidences his fascination with the philosophical system of P. D. Ouspensky, the Russian organicist, who saw organicism as an alternative to a scientifically mechanistic and economically determined way of looking at the world. Although Leopold expressed doubt as to whether organicism could be considered a literal truth, he described it as based on "intuitive perceptions, which may be truer than our science and less impeded by words than our philosophies."[3]

Leopold seemed to think of these alternatives quite explicitly as worldviews, although he never used the term. He, rather, referred to them as "conception[s] of the earth," one of which conceived the world as "our physical provider and abiding place," and the other conceived of a living earth, with "a soul, or consciousness," metabolism, and growth. But this growth was on too large a scale, and at too slow a pace to be experienced meaningfully by humans,[4] and so the organicist conception of the world seems less natural to us than does the mechanist one.

We obtain our metaphorical conception, it appears, without an explicit choice: "The very words *living thing*," he said, "have an inherited and arbitrary meaning derived from human perceptions of human affairs."[5] That is, our perceptions are conditioned by our actions, and our perceptions embody "arbitrarily" the metaphor characteristic of our activities. Leopold made clear that he doubted whether ultimate questions, such as the meaning of creation, could ever be demonstrated, either philosophically or scientifically, because "here again we encounter the insufficiency of words as symbols for realities."[6]

Thus Leopold first experienced the environmentalists' dilemma as a choice between two conceptions of the world. We can almost see the spirits of Pinchot and Muir peering over Leopold's shoulder in these passages: Pinchot, the practical man, set out to stop resource waste and build a "scientific" forest service. The earth to him, in his professional conception, was dead, manipulable, a source of products; and he spoke the language of economic determinism. Muir, the pantheist and mystic, saw the world as alive, and humans as a small part of a larger, dynamic system, their daily, economic activities representing an unavoidable blasphemy against a larger, ultimately spiritual whole. And Leopold, who was struggling to comprehend both viewpoints within a single management approach, must often have thought that his colleagues in the Forest Service and his allies from the preservationist camp[7] spoke entirely different languages.

If we left the matter here, we would have to conclude that Leopold had fallen into a form of relativism—one sees the world as alive, or one sees the world as dead. This difference is important because it affects the way we value and treat the earth, but acceptance of one or the other conception is an arbitrary result of the language we speak, and this in turn seems to follow without much thought upon our day-to-day concerns and activities. Pinchot, the professional forester, speaks "the language of compound interest," making him incapable of "seeing" aesthetic values. If, like Muir, we adopt a spiritualist rhetoric, we will be hard put to discuss wise-use issues.

But Leopold also suggested, in a parenthetical comment, a way out of the relativistic dilemma: "How happy a definition is that one of Hadley's which states, 'Truth is that which prevails in the long run'!"[8] Although Leopold never completed a list of references for this unpublished paper, he undoubtedly referred to the work of Arthur Twining Hadley, who was president of Yale University when Leopold was a student there. In a book that Leopold apparently read, Hadley said: "The criterion which shows whether a thing is right or wrong is its permanence. Survival is not merely the characteristic of right; it is the test of right."[9] This idea, which Hadley attributed to William James, provided Leopold with a way out of the relativistic conundrum: Human cultures, such as the dominant, economic exploitationism of American society, will be submitted to the test of time. If a culture uses up the land and extinguishes itself, it is refuted in the concrete world of history, even if it cannot be refuted in the abstract world of philosophy and systematic worldviews. This idea of applying a Darwinian criterion to cultures, their ideas, and their practices, advocated explicitly by Hadley, was no doubt attractive to Leopold, for it provided a neat unification of his biological and philosophical thought, by subsuming both under the Darwinian principle of "survival of the fittest."[10] Cultures, in their practices and associated ideas, must be adapted to their environmental con-

text. If they are sufficiently sensitive and dynamic—able to react successfully to changes in that environment—to survive for long periods, then
they have found "truth" in the pragmatic sense.

Rather than choose decisively between organicism and anthropocentrism, Leopold stated a common ground: "The essential thing for present
purposes is that both admit the interdependent functions of the elements."[11] Then, in the crucial point of the essay's argument, Leopold chose
to base his conservation ethic not on organicism, but on the shared consensus of interrelatedness, reasoning that organicism, to "most men of affairs"
is "too intangible to either accept or reject as a guide to human conduct."
Leopold thus followed Pinchot in seeking consensus in the arena of action.

Leopold immediately thereafter granted a human-centered system of value,
which he admitted was shared by Christianity and by scientific resource
managers, and continued to argue that the current treatment of the fragile
lands of the Southwest was immoral, even from the anthropocentric viewpoint.[12] By the "logic" of the anthropocentric viewpoint, there is a "special
nobility inherent in the human race," which implied to Leopold a responsibility to protect the interrelated functions of the larger system on which
our civilization depends, whether we are Christians, organicists, or scientific resource managers. The upshot, then, was not so much to advocate
any particular philosophical viewpoint, but to argue from a wide variety of
philosophical viewpoints that careless use of the land is immoral. If the
overall purpose of the essay was to develop an alternative to Pinchotist
management, it is clear why Leopold is so ecumenical and puts so much
weight on the opinions of "men of affairs," because he acted in the tradition of Pinchot's political consensus-building.

But Leopold also realized this task would not be as easy as Pinchot had
assumed. Pinchot had appealed to economic determinism and democratic development of resources to gain political clout, especially in the development-
oriented western states. But economic determinism and developmental
boosterism tempered only by concern for fair distribution, however successful politically, was bankrupt as a philosophy of land management. That
was obvious to Leopold from his observations of rampant erosion on the
fragile range of the Southwest Territories.

Leopold, therefore, saw his political task as one of forging a set of management principles that would be acceptable to mystical visionaries such as
Muir; to citizens, mainly Christians who never questioned human dominion over nature; and to his colleagues among scientific research managers.
He realized that, in addition to the relatively easy support gained for conservation that can be justified economically, he also needed a management
mandate for recognizing *limits* to the economic exploitation of nature. He
struggled to articulate a philosophy that would justify not only productivity,

but limits on productivity when modern methods of exploitation damage fragile ecological systems. Earlier in the essay, Leopold had offered the general scientific hypothesis that, for natural systems, "the degree of stability varies inversely to the aridity."[13] In the final, moralistic section of the essay, Leopold was searching for a set of shared principles adequate to limit destruction of the resource base, but that would also appeal to holders of a variety of worldviews.

Leopold the Manager

In 1920 Leopold traveled from Albuquerque, where he was the Forest Service's Director of Operations for the Southwest Territories, to New York to attend the Sixth American Game Conference. In a report on game conditions in the Southwest, he enthusiastically described the predator eradication program he had started there. Leopold had formed a coalition of sportsmen and stockmen to eliminate wolves, mountain lions, and other large predators from Arizona and New Mexico: "But the last one must be caught before the job can be called fully successful," he said. He was concerned that "no plans for game refuges or regulation will get us anywhere unless these lions are cleaned out," but was optimistic about the future: "When they are cleaned out," he said, "the productiveness of our proposed refuges and plans for regulation of kill, will be very greatly increased."[14]

In 1943 Leopold repented his war on wolves in a graceful and humble little essay written explicitly for the collection.[15] "Thinking Like a Mountain," Leopold's *mea culpa* on wolves, recounts his conversion from wolf eradicator to wolf protector. If we are to understand Leopold's land ethic, a good place to start is by comparing Leopold's early game management strategies with the end point of his wolf policy.

Leopold's early wolf-eradication policies were justifiable within the prevailing policy doctrines of the Forest Service, which expressed Pinchot's utilitarian approach to resource management. This is not surprising. Leopold obtained this master's degree in forestry at the Yale School of Forestry, which was begun by a large gift from Pinchot's family, and he entered Pinchot's Forest Service in 1909, immediately after graduation, when Pinchot was at the height of his influence over bureaucratic conservation policies.

Muir, as we have seen, railed against Pinchot's materialistic biases, and Leopold followed Muir in decrying the aesthetic sterility of management based purely on "ciphers." In his classic text *Game Management*, published in 1933, Leopold criticized artificial management of game to maximize "take," arguing that "it is not merely a *supply* of game, in the strictly quantitative sense, that is in question. The conservation movement seeks rather to maintain values in which *quality* and distribution matter quite as much as quantity."[16] Leopold also diverged from Pinchot's productivity-oriented model in developing and pushing through his proposal, in the early and mid-1920s, for the first wilderness area, the Gila, in the Gila and Datil National Forests. In defense of wilderness, he argued that the cultural advantages of retaining some ties to our wilderness roots outweigh and, in this case, justifiably limit the pursuit of material prosperity.[17] But such dissents were mainly registered as digressions from the main task, to maximize productivity in the National Forests.

Two further characteristics of Pinchot's scientific resource management were eventually forced upon Leopold's attention, however. Pinchot's approach tended toward *atomistic* management practices, and it assumed an essentially *static* set of background conditions. Both of these points require brief exposition.

Scientific resource management was atomistic in that it tended toward single-species management. Foresters preferred monocultural plantings of a single species of fast-growing and marketable trees. Game managers emphasized deer production and, while pheasants and other game might be encouraged as well, each game species had a "management plan" of its own. Pinchot's approach was atomistic in a second and related sense: management goals were pursued individually. In the first part of the 1923 essay discussed in the first section of this chapter, Leopold questioned this feature of Forest Service policy: "Take the Sapello watershed, which forms a major part of the GO's range in the Gila National Forest. Old settlers state that when they came to the country the Sapello was a beautiful trout stream lined with willows . . . but livestock has come in. And now the watercourse of the Sapello is a pile of boulders." Noting that that range was often cited as a shining example of range conservation, Leopold concluded, "The lesson is that under our peculiar Southwestern conditions, any grazing at all, no matter how moderate, is liable to overgraze and ruin the watercourses."[18] From this example, Leopold drew a generalization: "The effect of unwise range use on the range industry, or of unwise farming on the land, are all too obvious to require illustration. What we need to appreciate is how abuses in one of these industries in one place may unwittingly injure another industry in another place."[19]

So Leopold was quite aware of the problems caused by the atomistic bias of resource management practices prevalent in his day. It would be more than a decade, however, before he fully recognized the revolutionary implications of this insight for resource management. Only then did he develop an articulate explanation of its failings.

Leopold also came to realize that scientific resource management is based on an implicitly static model of biotic systems. It assumes that these systems are regulated by a "balance of nature." While this idea was usually explained with pat examples of predator/prey equilibria, and was often explicitly qualified as overly simple, it nevertheless had a profound effect on resource managers, including Leopold. It encouraged them to understand stability as *resilience*. The resilience of a system is measured as the rapidity and accuracy with which a system, once disturbed, will regain its predisturbance equilibrium.[20]

Any model that treats stability as simple resilience will necessarily be a static model—return to the predisturbance state is considered normal. Since scientific resource managers see natural systems mainly in terms of their productivity, resilience is closely related in the manager's mind with sustainability of yields of one or more target species. As long as the system remains in equilibrium, it will be assumed to produce a relatively consistent yield of the targeted species indefinitely. Or, to put the point the other way around, once system stability is thought of as resilience, stable yields of target species year after year are taken as an adequate measure of health of the productive system.

Leopold's early approach to deer management can be understood as an application of this idea of health and stability. Believing there was a natural equilibrium of the system that produces deer, Leopold saw wolves and mountain lions as constraints limiting annual production of deer for hunters. Removing wolves from the equation would provide more deer for hunters, thus increasing the yield of deer up to a maximum. Since hunters would substitute for natural predators in controlling populations of deer, a stable equilibrium could be maintained artificially. Indeed, Leopold apparently saw hunters as more controllable than other predators, hence his comment in the New York speech, quoted above, that once the wolves and lions were removed, "the productiveness of our proposed refuges and plans for regulation of kill, will be very greatly increased." Note that, in spite of his aesthetic concern that the hunting *experience* not be artificial, he was comfortable manipulating herds and interspecific predatory patterns to maximize deer for hunters.

Leopold was confident at this time that variations in regulations on hunting (such as choosing annually whether to have an open season on does) provided the manager with sufficient tools to regulate deer productivity. By managing only to maximize and sustain deer yields, Leopold acted atomistically; by assuming he could hold the system in artificial equilibrium, he acted on a static model. In this sense, Leopold applied, early in his career, Pinchot's atomistic and static management model.

Leopold's Transformation

We have seen that, as early as 1923, Leopold rejected the economic determinism of Pinchot, but he was unable to articulate a clear alternative. Then, in the late 1920s and early 1930s, Leopold fastened on three examples of failure in environmental management, and he played off them to articulate an alternative to Pinchot's management model. First, Leopold traveled to Germany to inspect German forestry. Second, population irruptions of deer were occurring on the very game fields from which he had helped to remove predators. And, third, Leopold observed the dust bowl and tried, by guiding the loving attempts of his family to reclaim a burnt-out and eroded farm on the Wisconsin River, to redress the effects of too-intensive farming on fragile soils.

In 1935 Leopold and five other U.S foresters were invited by the Schurz Foundation of Germany to visit Germany and to inspect and discuss new developments in German forestry. Leopold found the Germans, who had contributed so heavily to Pinchot's productivity-based policies, in retreat from a century-long commitment to monocultural plantings of spruce. Superior yields of spruce from even-aged, single-species stands planted in the early 1800s had encouraged the Germans to apply this policy pervasively.

After two or three generations, however, the outstanding yields were not repeated. Leopold summarized: "Litter failed to decay, piling upon the forest floor as a dry, sterile blanket which smothered all natural undergrowth, even moss. Roots ceased to penetrate the soil, lying in a tangled mat between the soil and the litter. . . . The topsoil developed excessive acidity, became bleached, and was separated from the subsoil by a dark band." Leopold noted that these conditions became known as "soil-sickness." "In short," he concluded, "pure spruce, the precocious child of timber famine and 'wood factory' economics, grew up into an unlovely and unproductive maturity."[21] The Germans, who Leopold credited with "teaching us to plant trees like cabbages," were retreating to a more natural approach, returning to mixed plantings including native species.

By 1935, however, results of his deer management policies had already forced Leopold to undertake a fundamental reconsideration of conservation policy: Wolfless game ranges were prone to irruptions in deer populations. During the same year that Leopold left his job as chief of operations for Southwest United States Forest Service and accepted a job as second-in-command at the United States Forest Service Forest Products Laboratory, biologists for the U.S. Biological Survey reported a runaway population of mule deer on the Kaibab Plateau. The range situation was critical, they said. Hunting, which had been ended on the Plateau when Roosevelt included it as part of the Grand Canyon Preserve in 1905, was reinstituted, but too late. Starvation caused a 60 percent reduction in the mule deer population during the hard winter of 1925–1926. This example involved both predator eradication *and* restriction of hunting, so it was debatable whether predator eradiction, hunting restrictions, or both caused the phenomenon. Leopold puzzled, from a distance, over what conclusions to draw.

He weakened his stance on predators, allowing in 1925 that "it is important to avoid the extermination of predators, but there is no danger of this as yet," and he continued to refer to predators as "varmints." Then, in 1927, one of his colleagues in the Southwest began warning of a similar population explosion in the central, less accessible portions of Leopold's own Gila Wilderness (where hunting had been allowed and even encouraged). In 1929, the trend toward overpopulation, especially of does, was clear; Leopold recommended that predator control efforts be shifted from mountain lions to coyotes. His former colleagues followed this advice but also, in a move that struck directly at the heart of Leopold's wilderness ideas, began plans to build roads into the Gila to increase accessibility for hunters.

After a trip to the truly "wild" area surrounding Rio Gavilan in Mexico, Leopold concluded that reasonable and stable populations of deer were possible without removing predators. He was greatly affected by this excursion, and later remarked that it was there that he "first clearly realized that land is an organism, that all my life I had seen only sick land, whereas here was a biota still in perfect aboriginal health. The term 'unspoiled wilderness' took on a new meaning."[22] Shortly after the trip, he asked his fellow conservationists whether "a normal complement of predators is not, at least in part, accountable for the absence of irruption [in the Sierra Madre]? If so, would not our rougher mountains be better off and might we not have more normalcy in our deer herds, if we let the wolves and lions come back in reasonable numbers?"[23]

Once Leopold undertook to complete the collection of essays that were to become *A Sand County Almanac*, his student and friend Albert Hochbaum goaded him into including an admission that he "had once despoiled." Pointing out that they had just killed the last lobo in Montana, Hochbaum said: "I think you'll have to admit you've got at least a drop of its blood on your hands."[24] The resulting essay, "Thinking Like a Mountain," came to be the centerpiece of Leopold's mature conception of human activity in the natural world.

Leopold mused on his journey from predator eradicator to land ethicist. He told how he and his fellow crew members shot a she-wolf from a distance, an incident that occurred on Leopold's first "reconnaissance mission" as a forester. Leopold approached the mortally wounded she-wolf in time to see "a fierce green fire dying in her eyes." "I was young then, and full of the trigger-itch," he confessed. "I thought that because fewer wolves meant more deer, that no wolves would mean hunters' paradise." Leopold perceived the result: the bones of the desired deer herd, "dead of its own too-much," littered the mountain.

The theme of the essay is time: "Only the mountain has lived long enough to listen objectively to the howl of a wolf," he said. Thinking like a mountain is putting oneself in the frame of time of the mountain, and "a mountain lives in mortal fear of its deer." And for good cause: "While a buck pulled down by wolves can be replaced in two or three years, a range pulled down by too many deer may fail of replacement in as many decades." Leopold was here recognizing how and why his deer management plan had gone astray: utilitarian, economically calculating management was management conducted from the time perspective of humans. Consumed with human cares, we strive for "peace in our time," and find it hard to see things from the mountain's viewpoint, a viewpoint measured in ecological and geological time, not human time.[25]

In order to help the reader understand the differing frames of time, Leopold the romantic created a double metaphor from his experiences. The mountain is personified: The dead sage becomes bones moldering, along with the deer bones, under the high-lined junipers. The mountain has not only a vegetative skeleton, though; it also thinks, and feels fear. If we are to manage nature without raising havoc, we must think as the mountain thinks.

There is another metaphor implied: The journey Leopold took from being a "game manager" to becoming a predator protector represents a symbol of the journey our culture must take in articulating and embracing a land ethic. And the metaphor of the living mountain drove the process; organicism aided him in thinking as the mountain thinks.

At about the same time Leopold the romantic wrote "Thinking Like a Mountain," Leopold the manager was trying to initiate a state-funded study of wolf ecology. He also drafted a more technical article entitled "The 1944 Game Situation." In it he stated his conclusions in a scientific tongue:

> It is probably no accident that the near-extinction of the timber wolf and the cougar was followed, in most big-game states, by a plague of excess deer and elk and the threatened extirpation of their winter browse foods. . . . It is all very well, in theory, to say that guns will regulate the deer, but no state has ever succeeded in regulating its deer herd satisfactorily by guns alone. Open seasons are a crude instrument, and usually kill either too many deer or too few.[26] The wolf is by comparison, a precision instrument; he regulates not only the number, but the distribution, of deer. In thickly settled counties we cannot have wolves, but in parts of the north [of Wisconsin] we can and should.[27]

The pattern of Leopold's thinking on wolf management was reinforced by his more detached observation of the dust bowl phenomenon. Leopold was careful to point out that he was not an agronomist and that his observations were only those of a nonprofessional. But he had traced what he called "illness" in natural communities, especially in the Southwest territories, even before venturing a general theory of fragility in arid systems in 1923, and was hence moving toward an explanation of the dust bowl well before it occurred. His explanation of the dust bowl was summarized in a succinct but powerful passage near the end of "Thinking Like a Mountain." Just after summarizing the lesson he had learned about mountains fearing deer, Leopold made the comparison "so also with cows. The cowman who cleans his range of wolves does not realize that he is taking over the wolf's job of trimming the herd to fit the range. He has not learned to think like a mountain. Hence we have dustbowls, and rivers washing the future into the sea."[28]

Leopold employed ecological terms to discuss his three case studies— German forestry, deer irruptions, and the dust bowl—and he thereby developed a paradigm of environmental management gone awry. He was ready, however tentatively, to articulate his criticisms of Pinchot-style management and propose a broad environmental management "theory" of his own. The new theory represented an application of Charles Elton's community model of ecological systems. Leopold met Elton in 1931 at a conference on natural cycles, and they became immediate friends. Already acquainted with Elton's important 1926 book, *Animal Ecology*, Leopold began in the 1930s to apply ecological theory to management problems.[29] According to Elton's theory, species are understood functionally, in terms of their contribution to a larger, biotic community, which Leopold understood as a layered pyramid: "The species of a layer are alike not in where they came from, or in what they look like, but rather in what they eat."[30]

Applying Elton's concepts, Leopold examined the evidence. The three case studies all resulted from attempts to increase the output of a resource base through more intensive management. In all three cases initial successes were registered as productivity increased dramatically. And in all three cases, following a few years or decades, a crash in productivity occurred. Leopold theorized as follows: "Each species, including ourselves, is a link in many food chains. A variety of species fulfill each function, each representing elements in many 'energy pathways.'" Contextual management conceives the context of management cells to be the larger, dynamic ecosystem in which the cell is embedded. When resource management activities simplify that system, it can become ill. This elaborated biota Leopold described as "a tangle of chains so complex as to seem disorderly, but when carefully examined the tangle is seen to be a highly organized structure. Its functioning depends on the cooperation and competition of all its diverse links."[31]

But human activities, if too "violent," can simplify such systems. Complexity, accumulated over millennia of evolution and competition, is reduced. The functions of the community—the flow of resources up and down the biotic pyramid—have been disrupted and made less efficient. The system is ill. Only sharp observers can see the gradual development of the illness, because it takes place in ecological time. But then, usually in response to stress—pest invasions, a hard winter, or a prolonged drought—the system collapses. It will not die entirely, but it may re-equilibrate at a lower level of functioning. In the worst case, the system goes into a tail-spin, as each downward turn in complexity results in lower productivity and vice versa, until the system has been reduced to a desert.

Based on this scientific interpretation of the three events, Leopold introduced the concept of "fragility" as applied to ecosystems. He also drew conceptually upon an early stint in the grazing office of the Forest Service's Southwest District and broadened the concept of "carrying capacity" to apply to all types of stress to natural systems caused by human use, including extractive uses. Leopold could now articulate and explain the senses in which the utilitarian approach to land management had failed. As we manipulate systems to produce maximal outputs of targeted species by interfering with normal ecological and evolutionary mechanisms, the system begins losing complexity. This impoverishment will not be obvious because it occurs slowly and has a latent stage in which the redundancy—the internal complexity of the system—is only gradually eroded. Once the redundancy of function is removed, the system is vulnerable to breakdowns in its energy pathways. The whole community can collapse. Elton's theory thus provided Leopold with a scientific vocabulary to explain both the collapse of productive systems under prolonged stress and also the latency of these collapses.

Leopold developed on this basis what I will call a "contextual" conception of environmental management in which each management problem and initiative should be considered on at least two levels, first as a *cell* and second as a *cell-in-context*. At first Leopold thought deer-wolf-hunters constituted a manageable cell in the system, but he learned that under the fragile conditions in the Southwest, they were not. He had also to pay attention to the context, "the mountain," the vegetative cover that gives structure, complexity, and a certain type of stability to ecological communities. Building on Elton's theoretical conception of a community, Leopold concluded that we must always manage any species *in its context*—that is, as a member of an ecological community. Management will not be considered adequate if (1) the activity decreases the overall complexity of the contextual system (causes "illness" in the community), or (2) if spillover effects of the activity disrupt productivity in a larger context (such as grazing destroying watercourses and trout fishing). Under these conditions, the larger system is in danger of destabilization (such as occurred on the deer-infested mountain).

Leopold's scientific, contextual model implied, then, a two-step process for forming environmental policy. It is permissible to manage cells according to maximization criteria, provided the maximization plan also passes a broader criterion: Economically motivated activities must not threaten the "health" of the larger context. In this respect Leopold's approach seems very similar to that developed more recently by the economist Talbot Page. Page argued that a wise materials policy would be guided by, first, a conservation criterion, which delimits a set of ecologically acceptable policies. A second criterion, which further discriminates the policy options that pass the conservation criterion, chooses that policy that will be most economically efficient.[32]

It seems accurate, therefore, to attribute to Leopold an *integrated* theory of management, one that recognized two levels. *Resource* management is reductionist and production-oriented, and follows a general approach of wise use of resource-producing subsystems; *environmental* management pays attention to the larger, autonomously functioning environmental context. Further, it employs an organic metaphor and an essentially medical model, using ecology as the analogue of physiology and arguing for enlightened management to protect community health. Resource management applies until productive activities threaten to cause rapid destabilization of their ecological context; then limits are invoked.

Despite these important theoretical breakthroughs, Leopold faced an impasse; his theory told him to manage contextually, taking into account the varied scales of time and space that interlock ecological communities. "Ecology remains the fusion point of the sciences," he said in 1939. And he gave credit to "economic biologists" for helping to build up the field of ecology. But ecology was not equal to its task, he feared:

> The emergence of ecology has placed the economic biologist in a peculiar dilemma: with one hand he points out the accumulated findings of his search for utility, or lack of utility, in this or that species, with the other he lifts the veil from a biota so complex, so conditioned by interwoven cooperations and competitions, that no man can say where utility begins or ends.[33]

Leopold's elegant theory was unfortunately impossible to apply, given the state of ecological theory in his day. Two weaknesses in the community model prohibited Leopold from giving the theory more precise application. First, the community model suffers from a difficult problem of "parts and wholes." What counts as a unit for ecological study? What counts as a unit for efficient management? The energy pyramid Leopold introduced represented only an abstract model; actual energy systems were so complex that ecologists differed about where to draw system boundaries. In some cases ecological systems seem quite naturally bounded, as in the case of a spring-fed lake with no outlets. But even in these optimal cases ecologists must caution that their boundaries are at best permeable membranes; they are better viewed as temporary conveniences rather than real, closed systems. Worse, most cases occurring in nature are infinitely more complex than lakes without outlets—in most cases habitats and microhabitats intertwine in an overlapping, inextricable complexity.

Second, while community ecology provided a general framework for conceiving species and complexes of species as embodied in a larger, contextual community, it lacked an adequate conception of community stability. Leopold's contextual theory required a dynamic conception of stability and ecosystem health, one capable of relating a resource-producing subsystem to its larger context, because the context is expected to change and develop as well. Stability conceived as resilience of the system, as it is often conceived even today, is too static too serve Leopold's new model. Lacking a serviceable conception of dynamic stability of larger systems in which management units are embedded, Leopold's management theory was disabled.

He recognized the problem clearly and argued persuasively, in 1939, that the equilibrium model was inadequate to the complexities of management. He warned that a static conception of balance of nature cannot successfully model natural systems: "To the ecological mind, balance of nature has merits and also defects. Its merits are that it conceives of a collective total, that it imputes some utility to all species, and that it implies oscillations when balance is disturbed. Its defects are that there is only one point at which balance occurs, and that balance is normally static."[34]

Lacking detailed information on the functioning of particular land systems, unable to designate system boudaries with confidence, and lacking an adequate theoretical notion for characterizing the stability and health of larger, environing systems, Leopold was stymied in his attempt to develop positive prescriptions based on ecology. He did the next best thing, by falling back upon his governing organicist metaphor and an analogical implication drawn from medicine: Lacking a cure for the disease, practice preventative medicine. Caution and humility should, he concluded, guide environmental managers.

Recognizing that humans are a part of the system they manage, Leopold drew a crucial distinction based on the "violence, rapidity, and scope" of human changes. Human changes differ from evolutionary changes, he said, which "are usually slow and local." Leopold therefore distinguished between management activities that are sufficiently violent to disturb their context and those that are not—the former interrupt the energy flows of the community in which they are embedded, and the latter do not. "The combined evidence of history and ecology seems to support one general deduction," he said, "the less violent the man-made changes, the greater the probability of successful readjustment in the pyramid."[35]

Despite difficulties of application, Leopold's central and correlative concepts of ecological fragility and ecological health were highly developed scientific constructs. He developed three parameters relating these concepts to human activities of various levels of violence. The extent of the simplification caused by human changes in ecosystem functioning varies according to three conditions:

1. *The Fragility Condition*. Every natural system has a certain degree of integrity, or self-determination, resulting from the accommodation that the indigenous species have made with the land and among themselves. Some systems, such as those in the arid Southwest Territories, are extremely fragile ("set on hair trigger"), and are extremely susceptible to deterioration in energy circuits as a result of overgrazing or other pervasive disturbances. They are prone to illness.

2. *The Population Condition*. Whatever the inherent integrity/fragility of a given system, larger human populations result in more pervasive changes and tend to cause greater modifications and simplifications ("greater violence").[36]

3. *The Time Scale Condition*. Natural systems have evolved in evolutionary and ecological scales of time, in which changes occur more slowly than do human changes (we do not yet "think like a mountain"). The greater the rapidity of human-induced changes, the more likely they are to destabilize the complex systems of nature.[37]

Given these parameters, I conclude that Leopold never questioned the *right* of humans to manage; he questioned, rather, our *ability* to manage, arguing that, given the present state of ecological knowledge and theory, we will often fail to foresee important effects of resource management initiatives. This generalization implies that management for human welfare must be limited, and that the limits must be scientifically determined through careful observation and controlled studies whenever possible. Leopold, at this point, had reached the limits of scientific knowledge regarding applied ecology, and he counseled preventative medicine. He saw environmental management as a race against change: Can we learn enough about how to manage and protect natural dynamic equilibria before we have so destabilized systems that there is no reversing the trend toward degradation?

Leopold's Legacy: The Land Ethic

As Leopold the manager questioned, in basic ways, the dominant methods of resource management, Leopold the romantic was also raising intellectually radical questions. He dreamed of a world in which human inhabitants would live in harmony with the natural world, a world in which the Mother Earth would be experienced as alive and soulful, a world in which

human society would limit its consumptive excesses willingly, out of love
and respect for the land on which it evolved and on which it continues to
depend, both physically and culturally. This would be a world in which the
extinction of a species would be mourned as a death in the family, and in
which whole ecological systems would be watched over and cared for like
living organisms in need of comfort and nurture. Leopold saw that this new
world, not here as yet but perhaps beginning to emerge as a new environmental conscience, would require changes in human consciousness. Leopold was not at all sure what form these changes would take, but he was
sure that it would be based on a recognition of interrelationships, and that
it must be sufficiently noble to value the future. A shift in governing metaphors would help, he thought, but the likelihood and exact direction of
the shift "is admittedly in a state of doubt and confusion."[38]

It was not at first easy for Leopold to hold his two personae in balance.
The romantic persona was strong in his early life, as he hunted and ram-
bled over the bluffs towering above the Mississippi River. Under the dis-
ciplined training of professional forestry, though, the romantic side was
sublimated in his early career—never entirely so, for we have seen that he
often questioned the grand scheme of utilitiarian resource management even
as he pursued the goal of using the land wisely for human purposes. And
in his personal life, in his hobbies, in raising his children, and in jotting
down notes and ideas in journals, he never lost sight of the sheer joy and
wonder that accompanied his early rambles.

Then, in the 1930s, as he saw Pinchot's grand plan for managing nature's
resources for human benefit crumbling, as he saw trout streams ruined by
upland grazing, deer starving on predator-free and overstocked ranges, and
insects devouring trees planted like cabbages, Leopold's romantic persona
reasserted itself in his work as well as in his personal life. While he remained a committed conservation leader, the great achievement of the last
decade of his life was the cobbling together of the essays that eventually
constituted *A Sand County Almanac and Sketches Here and There.* For
most readers, it is this collection of nature essays that assures Leopold a
lasting place in environmental thought and American letters. The book is
indeed the crowning achievement of Leopold the romantic; the question
worth examining, however, is whether the thoughts in *A Sand County Almanac* represent a final victory for the romantic Leopold over the manager,
or whether he achieved, in his last decade, a workable unification of the
environmental impulse. Did Leopold, who experienced within himself the
environmentalists' dilemma as it was bequeathed him by Muir and Pinchot, in the end opt for organicism and biocentrism, eschewing the manager's role? Or was the book an act of synthesis, pointing the way for future

environmentalists to escape their great dilemma? In the answer to these questions we will find the true nature of Leopold's legacy.

In the "Foreword," dated only six weeks before his death, Leopold explained that the essays in the third part of the book, which includes "The Land Ethic," "tell the company how it may get back in step." Leopold then lists three "concepts" as central to the book: "That land is a community is the basic concept of ecology, but that land is to be loved and respected is an extension of ethics. That land yields a cultural harvest is a fact long known, but latterly often forgotten." And, he says, "These essays attempt to weld these three concepts."[39]

It is the third aspect, the cultural harvest, that provides to Leopold the welding rods by which he joins ecology and ethics in a larger whole. The task of American culture, Leopold concluded, is to construct a mental model of the good life, but to do so within the constraints imposed by ecological systems. Conservation biology, the value-packed science, is an essential part of the search for a model of management. The cultural harvest from the land is the contribution of an organic conception of the good life. It is so because society does not yet have a definition of the good life that managers can use as a blueprint. In the tradition of Thoreau and Muir he connected the "quiet desperation" of materialism and consumerism with alienation from nature. And thus the task of the environmental manager, besides managing, is to play midwife to the public's ecologically informed idea of the good life.

In "Conservation Esthetic," the first essay in the final, "logical" section of *A Sand County Almanac*, Leopold states the problem of his profession, in which he and his colleagues strive "through countless conservation organizations to give the nature-seeking public what it wants, or to make it want what he has to give."[40] He then set out a gradation of outdoor recreation, from "trophy-hunting," indirect trophy-hunting (photography, for example), solitude, informed perception, and developing a sense of husbandry. "It would appear, in short," he said, "that the rudimentary grades of outdoor recreation consume their resource-base; the higher grades, at least to a degree, create their own satisfactions with little or no attrition of land or life." And, he concluded: "It is the expansion of transport without a corresponding growth of perception that threatens us with qualitative bankruptcy of the recreational process."[41]

And thus Leopold addressed the question Muir had shied away from: How can man in great numbers worship nature without destroying the natural qualities he worships? And how, more practically, can access to the wonders of wild nature be offered democratically without taming it? Leopold advocated access, but access carefully guided by ecologically trained

recreational guides. Enjoyment of nature must eventually become contemplative rather than consumptive. Teaching citizens to appreciate nature as
a land system is identified with the task of creating a culturally rich and
ecologically appropriate concept of the good life. "To promote perception
is the only truly creative part of recreational engineering," Leopold said.[42]
And the creative task of *A Sand County Almanac*, and of the publicly spirited environmental manager, to weld ecology and ethics with cultural welding rods, is none other than the task of creating for our culture, drawn
from many nations and cultures and facing staggering scientific and technological change, a definition of a good life in a good environment.

But Leopold was not foolishly optimistic; he recognized the difficulty of
such a task. In "Wildlife in American Culture," he surveyed the resources
available for constructing such a definition: "The culture of primitive peoples is often based on wildlife. . . . In civilized peoples the cultural base
shifts elsewhere, but the culture nevertheless retains part of its wild roots.
I here discuss the value of this wild rootage."[43] "There are cultural values
in the sports, customs, and experiences that renew contacts with wild things,"
Leopold said, "and these values are of three kinds:" (1) they stimulate
awareness of history; (2) they have the important value of "any experience
that reminds us of our dependency on the soil-plant-animal-man food chain,
and of the fundamental organization of the biota;" and (3) they encourage
the exercise of those "ethical restraints collectively called 'sportsmanship.' "[44]

Leopold hit the third point particularly hard: "A peculiar virtue in wildlife ethics is that the hunter ordinarily has no gallery to applaud or disapprove of his conduct. . . . Voluntary adherence to an ethical code elevates
the self-respect of the sportsman, but it should not be forgotten that voluntary disregard of the code degenerates and depraves him."[45]

Leopold also made clear that wildlife managers can, as indeed they must,
build the new cultural good with the building blocks that the existing culture offers—but he did not despair on that count. He found the cultural
resources he needed in the traditional American "split-rail values." "The
pioneer period gave birth to two ideas that are the essence of split-rail
value in outdoor sports. One is the 'go-light' idea, the other is the 'one-
bullet-one-buck' idea." These were, in their inception, "forced on us; we
made a virtue of necessity." But "in their later evolution," he noted, "they
became a code of sportsmanship, a self-imposed limitation on sport. On
them is based a distinctively American tradition of self-reliance, hardihood,
woodcraft, and marksmanship. These are intangibles, but they are not abstractions."[46] And these ideas, if they are not subverted by the sporting
press which "no longer represents sport" but "has turned billboard for the
gadgeteer," still provide the resources for a culturally viable conception of

outdoor recreation. In particular, if we remember that "outdoor recreations are essentially primitive, atavistic,"[47] the ability to turn necessity to virtue is yet available. It needs only, Leopold believed, to be nudged forward and defined for a new set of necessities implied by ecological science.

Wilderness preservation was given the same, cultural explanation. Wilderness embodies the historical context in which we struggle for durable values, and it "gives definition and meaning to the human enterprise."[48] And thus Leopold's organicism and contextualism expand to another level: Human culture and its values are a part of the living, growing organism as well. Humans, to know their place, metaphysically and morally, must preserve a healthy and dynamic system in which they can evolve morally. Restoration ecology, and the correlative task of developing a land ethic, as the odyssey of our civilization toward knowledge of ourselves as ecological, but rational, beings.

And thus the stage was set for the final piece, "The Land Ethic," easily the most discussed essay in conservation history. I will resist the temptation to summarize or quote extensively from that essay. I will rather offer some general observations that I hope will stimulate readers, who are encouraged to read the essay once again and to reconsider their interpretation of Leopold's "Land Ethic." Leopold is often cited as an advocate of a new morality, biospecies equality.[49] That Leopold saw new and grave responsibilities limiting human activities in the modern world of bulldozers and concrete is without question. But whether he saw these obligation as deriving from sources outside of, and independent of, human affairs seems to me doubtful.

The point may be mainly of interest to philosophers. But since there is a tendency, discussed in detail in the next chapter, to see modern environmentalists even today as divided, and divided primarily according to their acceptance or rejection of the moral hypothesis that nature has value intrinsically, it is at least worth examining whether Leopold did in fact posit such an independent value. I have emphasized, throughout my examination of Leopold's thought, that he preferred the organicist metaphor and that he believed this metaphor led to a deeper sense of love and respect for the land. Early in his career Leopold despaired of establishing, either rationally or in public debate, the literal truth of his guiding metaphor. And yet he promoted the organicist metaphor, using it as an informal guide to limit and inform his management experiments and, in his nature writing, as a means to improve the ecological perception of the American public. I see no evidence that Leopold, between 1923 and 1948, found some special means by which to prove the organic metaphor or its ethical implications literally true—true in some sense independent of culture. While there are passages in "The Land Ethic" that attribute rights to other species, we should remember that Leopold used philosophical concepts quite casually as tools to illuminate experience and should avoid taking such assertions literally.

On my reading, everything Leopold says in the "Land Ethic"—the references to ethics as "a limitation on freedom of action in the struggle for existence," to the land ethic as "an evolutionary possibility and an ecological necessity," to Ezekiel as a prophet whose time has now come, and, especially, to the rejections of "economic determinism"—can all be interpreted as guiding the search for a *culturally* defined value in nature. Moreover, these passages all find antecedents in Leopold's 1923 discussion in which he despaired of proving one worldview over another in any literal sense. Leopold rather saw the search for such an ethic as one culture's search for a workable, adaptive approach to living with the land. But Leopold was not dogmatic about the exact form of the ultimate values the culture would develop—the development of which, like the stability of the land organism, is a dynamic affair that builds on what exists and that changes in response to changes in the larger whole. Those changes, he increasingly recognized as his career unfolded, cannot be measured against some a priori truths, because philosophical pronouncements stretch the limits of literal truth; he increasingly measured them according to their results for management practices. If a society's practices are not adaptive to the ever-changing natural situation, the culture will fail. And so Leopold did not see himself as an advocate of nonanthropocentrism, a moral law he would dogmatically enforce on his fellow citizens, but as a fellow searcher after truth and a workable notion of the good life. Grand ideas get less play in Leopold's last "philosophical" statement—Ezekiel makes an appearance but Ouspensky and Hadley do not. Leopold was constructing moral and metaphysical principles out of improving management practices, rather than deducing the latter from the former. His method was broadly inductive, not deductive. Accordingly, he felt little compulsion to adopt a single value axiom.

Organicism appears briefly in the section "Land Health and the A-B Cleavage," but it is not presented as a grand philosophical system. Leopold saw it simply as a guiding metaphor for the group of conservation managers, "the dissenters," among whom he counted himself. This group, unlike group A, which "regards the land as soil, and its function as commodity-production," sees the land "as a biota, and its function as something broader."[50]

There is a sense in which my refusal to attribute to Leopold a full-blown, literal sense of independent moral values in nature is a philosopher's quibble. Surely, it could be replied, Leopold, if anyone, rejected the arrogance that derives from a narrow, self-serving anthropocentrism. And surely he did.

But my point in another sense is no quibble, for Leopold's decisive rejection of *narrow* anthropocentrism never, I believe, amounted to a dogmatic acceptance of the denial of human privilege in deciding what goals our policy should pursue. In the 1923 essay, Leopold was not only arguing that nonanthropocentrism would lead to better management practices, but also that Ezekiel (who was, of course, a prophet of anthropocentric Judaism), organicism, and long-sighted anthropocentrism (its sights lengthened by scientific observation) *all* point to the need for modified policies in the treatment of the fragile lands of the Southwest Territories. These three philosophies unite in opposition to economic determinism, which pays attention only to short-term profits. Organicism was attractive to Leopold in 1923, and in 1948, as an insightful and fruitful guiding metaphor. But the metaphor, without necessarily being established as literally or a priori true, can still guide policy. When it guides us, as in "Thinking Like a Mountain," to a more insightful interpretation of the scientific facts, it deepens that worth. And, when it focuses our values on systems and processes rather than outputs of a single species, it proves its worth. In the Darwinian sense, in which cultures have found the truth if their practices are adaptive and the culture survives, the organic metaphor can guide us to the truth. But this is not a truth independent of culture, a truth that can be used dogmatically as a stick to beat the culture into accepting moral imperatives independent of it. The metaphor, rather, guides a process—the cultural formation of a conception of the good life—in which conservation biologists have as much of a role as philosophers, but so also do cultural ideals such as freedom and independence and good sportmanship in hunting.

Leopold did not, I think, reject his professional role as manager of the biota. He rather recognized that, in the absence of an a priori, self-evident standard of what is right and good in management, the manager must develop a notion of the good life and simultaneously implement it. Organicism, while no more the self-evident standard than Pinchot's utilitarian formula, provides a useful governing metaphor. The intuitions that follow upon a soulful interpretation of the earth are sounder intuitions than those based on the interpretation of the earth as dead, and Leopold hoped that our culture would one day embrace organicism as its ruling metaphor. But failure among his colleagues and the public to make this leap of consciousness was not for Leopold a reason to write them out of the environmental struggle. Long-sighted anthropocentrists, from Ezekiel to moderns who accept the logic of nobility in the human enterprise, and organicists all should recognize the need to impose limitations on economic determinism.

The philosophical quibble therefore proves important practically: Leopold's lack of dogmatism in supporting his organicism precluded him from using it as a badge of acceptability, lacking which environmentalists must be described as "shallow," or not yet redeemed. He was willing to enlist all of the resources of our culture—from Ezekiel's Judaic prohibitions, to nobility supplemented with science, to "split-rail values" of pioneers—in order to construct a land ethic, a culturally viable ecological ethic. Cultural viability was essential to Leopold's activist task of building a political consensus behind an ecologically informed and contextually constrained approach to resource use. He was striving, therefore, for an inclusive, integrative ethic that could build on common denominators of many philosophies, not a dogmatic formulation of value that would exclude nonbelievers. We will see in Chapter 12 that some of his moralist successors have not always been so undogmatic.

By accepting a variety of bases to justify the environmental impulse, Leopold shifted the brunt of the search for a land ethic to science, away from economics on one side and from poetics on the other side. All of the moral systems that interact in our culture imply an obligation to the future to pass on an earth equally alive, and to do this, we must pay attention to the interrelationships as they dynamically change in response to our participation in the system. Ecology becomes a major guide in this task. We observe the effects of our actions, approve or disapprove them, and modify our efforts. Part of the task of the scientist is to define such terms as a "healthy ecosystem," "restoration," "natural area," "dynamic stability," and so forth.

Thus, Leopold's greatest legacy was to accept and elaborate Muir's view of ecstatic science, to reject the value-free scientific model of Pinchot, and to see conservation biology as a contributor to the construction of a viable conception of the good life. Leopold finally united his romantic and managerial personae by seeing the manager's role as that of helping the public to develop an ecological conscience; that ecological conscience must grow organically from our history, which is entwined with the evolution of our environment. If we wish these values to continue to grow organically, they must be worked out against the backdrop of natural necessities of evolution. To protect, even while guiding, that process is the greatest challenge modern humans face: Leopold recognized that we would need all of our cultural resources, ecological science, moral traditions, and pioneering spirit to forge a culturally viable conception of the good. In this sense, Leopold the romantic and Leopold the manager achieved a united persona in Leopold the naturalist.

4

Conservationists
and Preservationists Today

Death and Resurrection

Aldo Leopold died in 1948, of a heart attack suffered while fighting a brushfire that threatened the pines he and his family had planted at the shack. By dint of his strong personality, scientific curiosity, and near-universal respect from professional colleagues and nonprofessional wilderness advocates alike, Leopold had personified the search for a unified vision to guide human use of the land and offered a unifying model for conservationists. His death, at the height of his intellectual and leadership powers,[1] left the movement effectively rudderless.

The country was at the time being swept away by the postwar economic boom and a new period of unrestrained economic growth pushed environmental concerns off the political agenda. Stephen Rauschenbush, writing in 1952, said, "Conservation is in danger of becoming a lost cause" and provided a list of five "forces and events that had battered away at the old ideal."[2] Grant McConnell similarly described the conservation movement in 1954 as "small, divided and frequently uncertain." Wise-use conservationists and land preservationists appeared in the eyes of their opponents, McConnell lamented, as "but the representatives of particular interest groups and . . . hence no better than those whom they accuse."[3]

But the same economic growth that placed conservationists on the defensive in the late 1940s and 1950s created a new demographic situation, including a baby boom that resulted eventually in a large surge in the youthful segment of the population. Increasing incomes allowed discretionary spending on amenities. Outdoor recreation became more popular and vacation homes near the shore or in the mountains proliferated as more families chose to spend time in areas with natural amenities. Environmental historian Samuel Hays explains the resulting shift as focusing interest and concern increasingly on the quality of life. New social, economic, and demographic factors created a context in which the older conservationist concerns about efficient development and use of material resources—concerns with production—were replaced, in the postwar era, with increasing concern with the consumption side of the picture.[4]

These changes created a new constituency for environmental groups. Equally important, the complacent attitude of the 1950s was transformed by a new social consciousness in the 1960s. Although early protests focused on civil rights and protests against the war in Vietnam, the tendency to question and criticize the status quo soon brought to public attention a new set of more complex environmental problems.[5] These "second-generation environmental problems," as they have been characterized by Robert Mitchell, differ from earlier problems faced by either the conservationists or the preservationists. First-generation environmental issues typically involved a threat to a particular area or species. In the 1950s and 1960s, however, environmental problems emerged that were pervasive, delayed, and subtle. Air and water pollution, the proliferation of chemical wastes, and widespread use of pesticides brought about significant threats to human health and to entire ecological systems. The delayed effects of activities causing these problems made conclusive demonstration of threats difficult and made attempts to correct them more expensive, sometimes threatening entire industries or sectors of the economy.[6] At the same time, these pervasive problems struck directly at the amenity values that had become increasingly important to the enlarged and more affluent, consumption-oriented constituencies that had emerged in the postwar era.

In 1962 Rachel Carson published *Silent Spring*. The book, which loosened the floodgates of tacit concern, sold 100,000 copies in the first three months of publication, and focused attention on problems of pesticide use, shifting the focus of environmentalism to the more pervasive second-generation environmental problems.

Carson was trained as a zoologist and served as a career biologist with the U.S. Bureau of Fisheries and the Fish and Wildlife Service. Drawn since childhood to the sea, she led her generation in popularizing oceanography through her most characteristic works, *Under the Sea Wind* (1941), *The Sea Around Us* (1951), and *The Edge of the Sea* (1955). Carson was deeply imbued with an ecological view of the world and used it in these early works, not polemically, but to encourage humility as a human response to the natural world.[7] She believed that the ecological viewpoint undermined the arrogant attitude that caused humans to try to gain full control over their natural surroundings, and praised Albert Schweitzer's philosophy of reverence for all life.

Responding to the complaints of a friend in New England who watched songbirds die agonizing deaths in her yard after aerial spraying of DDT, Carson recognized the need for an all-out polemical attack. *Silent Spring* could not be like any of her other books; whereas they had been unpolemical and promoted ecological perception for its own sake, the new book had to be polemical and use ecological thinking for unabashedly political ends. Whereas in her earlier writings she had promoted a nonanthropocentric value position, her new task became one of establishing a threat to human health and survival. As Donald Fleming says, "Rachel Carson had to calculate her effects as never before, trim her sails to catch the prevailing winds."[8]

Other influential publications followed, including Barry Commoner's *The Closing Circle*, a more general indictment of technological arrogance, a series of books announcing a population crisis, and influential computer models projecting worldwide resource crises.[9] The net effect of these events, which intensified the more general demographic trends described above, was to expand greatly the constituency to which environmental groups could appeal, resulting in a rapid expansion of the membership of the traditional conservation organizations and a proliferation of new groups.[10]

The upsurge in environmental concern culminated in Earth Day 1970, which marked a peak of public concern for environmental issues in the modern era.[11] A conspicuous absence of polls testing public opinion on environmental issues before the 1970s suggests the unimportant role they played in national politics.[12] A 1964 survey by Louis Harris found that rising public interest in the problems of air and water pollution was the most frequently recurring theme.[13] In 1965, a Gallup poll found that 17 percent of Americans placed "reducing pollution of air and water among the top three problems the government should devote most of its attention to in the next year or so."[14] Gallup asked the same question immediately after Earth Day 1970 and "reducing pollution of air and water" was chosen by 53 percent of respondents. When this question was repeated in 1980, the response had leveled off at 24 percent, a net increase of 7 percent over 1965. In-depth questions, furthermore, show that while concern for environmental protection was conceived less as a crisis situation in 1980 (thus explaining its choice by fewer respondents as one of the three most urgent problems for government to attend to), support for national environmental protection efforts remained strong.[15]

More recent data suggest a resurgence of environmental concern in the general population, which is partly seen as a reaction to the Exxon oil spill in Alaska. According to a 1989 Gallup poll, three-fourths of Americans now think of themselves as environmentalists, and almost half of these think of themselves as "strong environmentalists."[16] A reaction to the Reagan Administration's turn away from environmental objectives resulted in a windfall of new members for environmental groups—the Sierra Club, for example, increased its membership from 180,000 in 1980 to 335,000 by early 1983—and a resounding repudiation of Reagan's policies in the polls.[17] The strength of support for environmental causes is indicated by polling data showing that 74 percent of Americans agree that environmental improvements must be made regardless of cost,[18] and that 92 percent of citizens rank "protecting the environment" as an important goal, whereas only 82 percent so rank making "progress on the control and elimination of nuclear weapons"![19]

Popular opinion, to summarize, shows very strong and general concern for environmental issues, but considerable variation exists across both time and place regarding the intensity and priority given environmental goals. International crises (such as the 1973 oil embargo) and triumphs can push environmental issues off the agenda temporarily; local areas that are economically depressed tend to favor economic growth over environmental protection.[20] What is unquestionable today, however, is the unwillingness of citizens to weaken current environmental programs.

Paradigms and Dichotomies

Speaking historically, it has been useful to characterize two competing and often opposed *groups* of environmentalists clustered around the banners held high by Muir and Pinchot, respectively. In this section, several attempts to discriminate environmentalists into opposed groups will be examined. Does the positing of two opposed and largely exclusive groups of environmentalists help to describe and explain the current situation in environmental politics?

Commentators on enviromentalism apparently think so. Even today the literature on the movement often refers to two factions of environmentalists and, while labels often vary, their pedigree is made clear by explanatory invocations of the original split between Muir and Pinchot. For example, Stephen Fox's stellar historical analysis of the environmental movement refers to amateurs ("in the tradition of John Muir") as the "driving force in conservation history" and describes these amateurs as "the movement's conscience." These he contrasts with the professional conservationists and governmental agencies, on which the movement depended "for expertise, staying power, organization, and money."[21]

The philosopher John Passmore, by contrast, distinguishes conservationists from preservationists not according to professional status but according to commitment to apparently conflicting motives.[22] Passmore devotes consecutive chapters to "Conservation" and to "Preservation," beginning the former with definitions: "To conserve is to save. . . . I shall use the word to cover only the saving of natural resources for later consumption." He continues: "Where the saving is primarily a saving *from* rather than a saving *for*, the saving of species and wildernesses from damage and destruction, I shall speak, rather, of 'preservation.' "[23] Passmore notes that conservationists and preservationists often work together on particular issues, but "their motives are quite different."[24] He later delineates the crucial distinction in motives by noting that, whereas conservationists' motives for protecting species and ecosystems can be understood in terms of instrumental values to humans, the preservationists' motives must be understood in terms independent of human interests: "As it is sometimes put, they [wildernesses and species] have a 'right to exist.' "[25]

In a more tendentious mode, "deep ecologists" contrast their own approach to environmentalism with another, less radical approach that they originally referred to as "shallow ecology." Due to the Scandinavian philosopher Arne Naess, this categorization is normally drawn according to clusters of beliefs and axioms, which function somewhat similarly to what I have called worldviews. Naess first defines the shallow ecology movement as seeking a single objective: "the health and affluence of people in the developed countries," and contrasts this with the deep ecology movement, characterized by adherence to seven principles:

1. rejection of the man-in-environment image in favor of *the relational, total-field image;*
2. biospherical egalitarianism;
3. advocacy of diversity and symbiosis;
4. anti-class posture;
5. opposition to pollution and resource depletion;
6. complexity, not complication; and
7. local autonomy and decentralization.[26]

At first glance, acceptance or rejection of this complex list of characteristics would hardly seem likely to define exclusive groups: How are we to categorize environmentalists who accept only some of the seven principles? Naess responds that it is the "global character" of deep ecology, "not preciseness in detail," that separates it from shallow ecology—that is, there will be differences among deep ecologists but, I think he means to say, they are united by accepting a general framework of beliefs, a systematic worldview, as a single, monolithic piece.[27] Naess's claim that there is a distinct breed of environmentalist—deep ecologists—apparently assumes an empirical hypothesis: Environmentalists adopt a new worldview all at once; they do not change gradually, altering some of their assumptions while maintaining others.

Another defender of deep ecology, Warwick Fox, responds to this problem by treating the labels for clusters of beliefs/principles as representing "ideal types" that are put forward as "heuristically useful." He therefore does not consider the categories exclusive or mutually exhaustive of the positions environmentalists might defend.[28] Fox's approach, then, treats these terms much as we have treated the terms Moralist and Aggregator—as shifting labels for ideal arguers, rather than as belief systems characterizing well-defined groups of environmentalists.

Whichever model is used, deep ecology is defined in contrast to shallow ecology, and shallow ecology is defined as human-centered; Naess's principle (2), biospherical egalitarianism, appears to be the key. An environmentalist should be classified as *deep* if he or she believes nonhuman elements of nature have independent value.[29] It is characteristic of social scientific descriptions of environmentalism to designate environmentalists according to the acceptance of one or the other of two "paradigms," the dominant, human-oriented, social paradigm (DSP) and a new, biocentric, environmental paradigm (NEP).[30] We can examine the descriptive usefulness of positing exclusive groups of environmentalists by evaluating empirical work, mainly by sociologists, which employs the paradigm construct.

In some cases paradigmatic analysis is combined with the deep ecologists' idea that certain constellations of beliefs—"systems of global character" in Naess's terminology—serve to categorize and motivate two distinct groups of environmentalists. For example, the sociologist Bill Devall, citing Naess, describes "reformist environmentalism" (which he prefers to the blatantly pejorative "shallow ecology") as accepting the dominant paradigm in North America. A paradigm he defines as "a shorthand description of the world view, the collection of values, beliefs, habits, and norms which form the frame of reference of a collectivity of people."[31] Devall then defines "reformist environmentalism" as referring "to several social movements which are related in that the goal of all of them is to change society for 'better living' without attacking the premises of the dominant social paradigm."[32] Deep ecologists, according to the essential characteristics attributed to them by Devall and George Sessions, reject human dominance over nature, believing that nature has intrinsic worth; their belief in "biospecies equality" puts them outside the dominant social paradigm and distinguishes them from reform environmentalists.[33] Acceptance or rejection of the dominant social paradigm, therefore, functions as a characterization and explanation of the division of environmentalists into two distinct factions.

Note that this sociological interpretation of the current situation appears to represent an empirical theory to explain the behavior of environmentalists: Environmentalists fall into two opposed camps because they are polarized by commitments to two opposed paradigms. This explanation *presupposes* that environmentalists in fact fall into two camps—a sociological hypothesis deserving, certainly, of empirical test. Devall, however, presents no data to support this presupposition. Lacking such empirical data, the proposed explanation is circular; it is no more than a relabeling of the disjoint motive-sets that Passmore attributes to conservationists and preservationists: Reformist environmentalists pursue anthropocentric utilitarianism, while deep ecologists are nonanthropocentric.

We can now see the importance of the difference, exemplified by Naess and Warwick Fox, between treating the labels "deep ecologist," "shallow ecologist," or "reform environmentalist" as, with Naess, labels for systems of thought adopted monolithically and as, with Fox, ideal types that may or may not correspond to well-defined, exclusive groups of real environmentalists. Devall, playing on Naess's assumption that paradigms are characterized by some finite number of principles that will be accepted or rejected monolithically, sees no need to show that environmentalists array themselves in two opposed groups. If paradigms are opposed systems of thought that must be accepted monolithically, then it follows that, if they are to avoid self-contradiction, environmentalists who choose one system or the other will form exclusive groups.

The notion of a paradigm was first introduced by Stephen Toulmin and Thomas Kuhn to refer to the shared systematic set of assumptions and methods that give unity to a scientific discipline and define an acceptable explanation of phenomena studied by that discipline.[34] It is important to note, however, that scientific paradigms can be argued to be either "closed" or "open." Kuhn went beyond Toulmin in arguing, essentially, that scientific paradigms were *closed:* That is, the basic assumptions of a Kuhnian paradigm apparently serve to limit the "normal" discourse of the discipline's practitioners.[35] The questions that are raised and answered are given form by the disciplinary principles constituting the paradigm. A question that cannot be framed in those terms is excluded from the discourse of the discipline—even to pose such a question would require a paradigmatic "revolution."

Debate has raged, since the original publication of Kuhn's influential book *The Structure of Scientific Revolutions*, over whether scientific paradigms are in fact closed, as Kuhn claims. Whatever is decided on that important issue in the philosophy of science, it seems without question that environmentalists and other ordinary people approach problems of their political life—such as what policies to support and what candidates to vote for—using open systems of thought. Most of us, most of the time, operate with open systems of belief. Unquestioned axiom sets are often incomplete—they do not determine all decisions in any direct sense—and we probably lack deductive validation for most decisions. Environmental historians, sociologists, and survey researchers, nevertheless, have increasingly used the concept of a paradigm to refer, more broadly, to the constellation of " 'common values, beliefs, and shared wisdom about physical and social environments' which constitute *a society's* basic 'worldview.' "[36]

Labels such as "deep ecologist" or Passmore's "conservationist" and "preservationist" are given meaning within systems developed by philosophers. Because they are explicitly concerned with the justification of beliefs and actions, philosophers construct systems that are as highly connected as possible (various parts of the system are linked by strong argumentative connections). Because of this strong connectivity, these relatively closed systems tend to be accepted or rejected monolithically, and in this respect paradigms do not serve as useful models for actual systems of belief or worldviews such as those that guide active environmentalists.

Philosophers' idealized systems are therefore best thought of as having a prescriptive function—they provide a template for evaluating various rationales and styles of argument for environmental policy objectives. It does not follow that these idealized systems will be descriptively useful, because most activist environmentalists will operate, when they consider and discuss policy, with systems that are not closed, as are the idealized systems proposed by Naess and Passmore. And, because they operate with relatively open systems, environmental activists are less likely to adopt and reject systems of thought monolithically—they are more likely to adjust and alter their beliefs and attitudes gradually. When social scientists borrow relatively closed systems of thought designed for prescriptive purposes and treat them as categories descriptive of factions of environmentalists, they risk forcing activists into categories that are far more rigid than the data warrant.

For example, when the sociologist Devall sees two factions of environmentalists clustering around Naess's two systems of thought, he implicitly assumes that the belief systems of environmentalists are relatively closed systems—sets of axioms and principles that are accepted or rejected monolithically. But this assumption excuses him from the real empirical task, to show that environmentalists, in their actual behaviors—not in the rhetoric that they use to explain those behaviors—act as two exclusive and opposed groups.

Smoke and Mirrors: The Empirical Evidence
for a Dichotomy among Environmentalists

An explanatory theory of why two groups act differently, such as the theory that environmentalists behave as members of two opposed groups, is only as good as the empirical evidence for opposed behaviors. What, exactly, is the empirical evidence that environmentalists are indeed arrayed into opposed camps today? So far we have concluded that emphasis on *closed* systems—paradigms—of opposed principles has encouraged some sociologists to skip the essential step of empirically supporting the view that there are, in fact, two *groups*, characterized by acceptance of opposed value systems and given to acting in opposition to each other. Now we must consider some of the evidence that has actually been cited in support of that opposition.

Immediately, however, we are faced with a confusing ambiguity. Whereas Devall uses the concept of a new and deeper ecological paradigm to explain a difference among two groups of environmentalists, Riley Dunlap and Kent Van Liere use *the same distinction* to explain why some members of *the general population* show lack of environmental concern while others, environmentalists, support environmental causes.[37] Dunlap and Van Liere hypothesize that "individual commitment to the dominant social paradigm is *negatively* related to environmental concern," and found that their data from 806 mailed questionnaires strongly confirmed the hypothesis.[38] Here, the concept of a dominant social paradigm is used to clarify and explain the difference between environmentalists and nonenvironmentalists: Those who express concern for the environment and work to promote environmental causes are shown to differ in paradigm commitment from those who accept the dominant social paradigm. Those who accept the dominant social paradigm, by contrast, show little concern for environmental problems.

It might justifiably be objected that commitment to, and rejection of, the dominant social paradigm is being asked to carry a great deal of explanatory weight, embracing as it does two quite different dichotomies. The ambiguity also creates a situation ripe for confusion. An example of the expected confusion appears in Milbrath's reports of the results of a major, three-nation study of environmentalism, administered in 1980 and 1982.[39] Milbrath distinguishes between an "Old Dominant Social Paradigm" and a "New Environmental Paradigm," according to acceptance of six components. The first component concerns basic values: whereas proponents of the DSP value nature to produce goods, acceptance of the NEP is marked by placing a "high valuation" on nature, explained as involving three subcomponents: Nature is valued for its own sake, humans should live harmoniously with nature, and environmental protection is preferred over economic growth.[40]

These opposed paradigms are then used in the interpretive sections of the book to express Milbrath's central hypothesis that environmentalists are becoming a "vanguard" of social activism that may become a force toward radical social change.

Milbrath selected a group of environmentalists, drawing his sample from membership lists of U.S. environmental organizations,[41] and compared their views with those of other "elite" groups (such as business leaders, labor leaders, and elected officials). Milbrath, therefore, chose a group that he took as representative of U.S. *environmentalists as a whole* and tested whether they accepted a new paradigm involving acceptance of the value position usually cited as characteristic of deep ecologists.

Milbrath's data are interesting and important (if not surprising). He found, for example, that 62 percent of the general public favored (with varying degrees of strength) a society that emphasizes environmental protection over economic growth, that 84 percent of environmentalists shared that belief, but that only 29 percent of business leaders agreed. This finding justifies the conclusion that environmentalists' views on economic policy differ significantly from those of business leaders and the general public, and may add support to the hypothesis of Dunlap and Van Liere, even though importantly different definitions of the paradigms were used in the two studies. Responses to value/attitudinal questions also revealed significant differences: For example, 78 percent of environmentalists but only 40 percent of the general public and 24 percent of business leaders preferred a society that preserves nature for its own sake. Again, this finding is used by Milbrath to illuminate the interesting differences between environmentalists and other members of the general public.

But there is something odd about Milbrath's presentation and interpretation of his data. The central idea of the interpretive sections of his book— that there is emerging a "vanguard" of environmentalists, a faction whose members accept a new paradigm (defined in heavily value-oriented terms) and are more dynamic in working toward basic changes in social and especially environmental policy—floats entirely free from the volumes of data he presents. Those data were derived from questionnaires designed to compare the views of environmentalists with those of other members of society. Nothing in the data, as presented and aggregated in the detailed appendices concluding the book, provides evidence that the subset of environmentalists Milbrath designates as vanguard-members actually represents the same environmentalists, the 78 percent, who prefer a society that preserves nature for its own sake. To show this, Milbrath would have to present more finely textured data comparing the value commitments and action programs of the subgroups of environmentalists. No data are presented to show that these groups are cohesive or that they are best understood as opposed rather than as stretched along a continuum.

And yet Milbrath has a good deal to say about these subgroups. He mentions deep ecologists, noting that they represent a small subgroup of environmentalists, and says that they "do not constitute a strong force for near-term social change."[42] He also recognizes another group, which he calls "nature conservationists," the members of which "place a high valuation on a safe and clean environment but tend also to adhere to many of the beliefs, values, and social structures of modern industrial society."[43]

Milbrath's generalizations about environmentalists as a whole seem to conflict with what he says about their subcategories. He concludes, on the one hand, that "these modern-day prophets, these environmentalists, are beginning to develop a new environmental paradigm."[44] At another point, he says: "*Nearly all* of the environmentalists believe that nature should be preserved for its own sake."[45] What, then, are we to make of Milbrath's category of "nature conservationists"? They would seem to represent counterevidence against his hypothesis that environmentalists as a whole are adopting a new paradigm, and yet Milbrath apparently accepts them as environmentalists but not as vanguard environmentalists.

Milbrath is, apparently, aware of the disparity between his data and conclusions, because he devotes a brief paragraph to the mention of some finer-textured correlations he ran on some subsets of the data. Charitably put, this brief discussion hardly supports the conceptual edifice of subcategories he imposes on environmentalists as a whole.[46] The reader suspects that Milbrath's subcategories of environmentalists are more reflections of his conceptualizations and interpretation than empirically grounded categories.[47] Whereas he presents data to support the conclusion that environmentalists share a different paradigm than do business leaders, he *defines* the subgroups of environmentalists according to acceptance or rejection of a philosophical construct.

There is another reason to question the reality Milbrath assumes for subcategories of environmentalists. He makes much of the fact that 78 percent of the environmentalists he questioned prefer a society that emphasizes preserving nature for its own sake over a society that emphasizes using nature to produce the goods we use. He concludes that most environmentalists reject anthropocentric values. If this is so, we might conclude, their values will be similar to those of deep ecologists as defined by Devall and Sessions. But Milbrath also presented the environmentalists with a chance to state preferences between a society that saves its resources to benefit future generations and a society that uses its resources to benefit the present generation. Here, 88 percent of environmentalists chose a society with concern for future generations. If 78 percent of environmentalists accept nonanthropocentrism, then the question on future generations should have been at least troubling for them—to treat natural objects as a resource for human use (either present or future) is to adopt the very anthropocentric attitude that deep ecologists and, from Milbrath's data, 78 percent of environmentalists reject. If environmentalists have accepted a new, biocentric paradigm, they have apparently done so without much concern for consistency.

What has gone wrong here? Consider again the question on nature for nature's sake. Given a choice between agreeing strongly, moderately, or weakly with the assertion that one prefers a society that preserves nature for its own sake or a society that uses nature to produce the goods we use, self-affirming environmentalists will lean toward the clearly environmental option. Thus it is not surprising that Milbrath found only 8 percent of environmentalists who chose one of the productivity options. The problem with this question is that it allows no response to the effect that the individual would prefer a society that emphasizes saving nature for human but noncommercial amenity uses, such as hiking, viewing, naturalist study, or spiritual uplift. If such responses had been possible, in addition to the two offered, Milbrath would no doubt have found a multinodal distribution of environmentalists who reject a society that uses nature only to produce the goods we use. When Milbrath found two factions of environmentalists, those who accept and those who reject the anthropocentrism he associates with the DSP, he saw a reflection of the false dichotomy he offered his respondents.

The moral of this story is, I think, that acceptance of particular paradigms and other categories developed by philosophers, such as "deep ecologist," are not necessarily useful for the descriptive purposes of social scientists. As noted in Chapter 1, when I asked activist environmental leaders whether they found the labels "conservationist" and "preservationist" useful today, they almost all responded that these labels have meaning, at most, as pejoratives, referring to those too willing to compromise with developers and those who want to "lock up" resources, respectively. Trying to describe environmentalists as arrayed in two camps according to their acceptance of differing value systems and associated closed paradigms has led to imprecise use of data and confused explanations of how environmentalists differ from nonenvironmentalists and from each other.

I propose another approach: Suppose "conservation" and "preservation" are taken not as labels to distinguish *groups* of environmentalists who are distinguished by commitment to opposed *motives*, but as two different *activities*, corresponding roughly with the original dominant concerns of Pinchot and Muir. This change in approach reduces the importance of commitment to diverging values and attendant explanations in terms of divergent motives, and instead focuses attention on patterns of individual behavior in specific situations. *To conserve* a resource or the productive potential of a resource-generating system is to use it wisely, with the goal of maintaining its future availability or productivity. *To preserve* is to protect an ecosystem or a species, to the extent possible, from the disruptions attendant upon it from human use of the species or its habitat. If one then insists on classifying environmentalists, one can do so derivatively and in more-or-less terms: Conservationists and preservationists, respectively, are individuals who usually act in one way or the other.[48] This usage should not be taken, however, to imply that there are two solid groups always arrayed in opposition. The categories represent tendencies to emphasize, in various situations, one rather than another point on a spectrum of shared values and concerns.

5

Worldviews:
A Whirlwind Tour

The Worldviews of Environmentalists

Languages are more like searchlights than floodlights; they do not illuminate equally across the full range of the perceptual field. This generalization is especially true of technical vocabularies that characterize specialized professions such as economics and ecology, and to a lesser extent of systems of favored expressions used by groups that share a special interest, such as nature preservation. Because of this selectivity of focus, languages and special vocabularies can disclose the world differently.

If various environmentalists' descriptions of the same events often sound as though they are perceiving something quite different, it is because environmentalists, who have not adopted a shared worldview, express themselves in a variety of unsystematized languages and vocabularies. It will therefore be useful to examine the systems of concepts that important environmentalists have used to describe what they have perceived.

It is important not to think of concepts as abstractions, however. A language and the forms of expression it embodies are the glue that hold worldviews together; our vocabularies therefore carry our perceptions, playing an active role in shaping the world we see. The very meaning attributed to events is affected by the interplay of perceptions, actions, and the linguistic behaviors that express the meaning of events.

Perceptions, we assume, directly affect the theoretical hypotheses and conjectures we develop to make sense of our world as we act within it. But theoretical assumptions likewise affect perception; and since perception is the only basis we have for discriminating among theories of reality, it is difficult to avoid the conclusion that to some degree at least, the constellation of conceptual, theoretical, and value precepts we operate with, and the vocabulary we use to express them, will determine the shape of the world we encounter.[1] This, loosely put, is the idea behind worldview analysis, as introduced in Chapter 2 and used occasionally throughout this book. Because most people are not accustomed to question their linguistic forms and the philosophical commitments embodied in them, it is often useful to discuss these issues informally in metaphorical terms. For example, the language of atomism as applied to nature is often associated with a mechanistic metaphor, while more holistic approaches are associated with an organicist metaphor. That is, comparing divergent metaphors often represents an attempt to formulate basic choices regarding worldviews.

A worldview, as we are using the term here, refers to the constellation of beliefs, values, and concepts that give shape and meaning to the world a person experiences and acts within. A worldview is not necessarily a well-developed, systematic philosophy. It can be, but the worldviews of most people remain simply sets of background assumptions, often not even recognized by those people, against which they understand the world and act in it. Individuals often act on unsystematized and incomplete conceptions, on fragments of worldviews.

Even when a worldview is not explicitly systematized, it may nevertheless prove useful to identify certain principles as most central and formative—we can refer to these as "axioms." The axioms of a worldview, while often inexplicit and hidden, represent rock-bottom commitments that the holders of that worldview would eventually cite as supporting the larger edifice of their beliefs, if they devoted adequate time and effort to the task of organizing their beliefs into a well-ordered system.[2]

Worldviews are best understood as action-oriented. While particular principles and elements of a worldview can be isolated, discussed, and evaluated abstractly, the worldview itself derives its character from its ability to guide action. This interpretation of worldviews is important because it ensures that we will recognize that all people, not just abstract thinkers who discuss their principles publicly, have a worldview. That the formative constituents of a worldview are often tacit, and not explicitly expressed or justified, does not detract from their essential role as a guide to action.

So that we might be more systematic in our evaluation of environmentalists' worldviews, we can provide four criteria, or "scales," for evaluating these constructs.

1. *The Scale of Completeness.* Do the elements of the worldview provide a basis for interpreting all facts and do they provide guidance regarding policies to cover every possible situation?
2. *The Scale of Connectedness.* To what extent are the various elements

of the system derived from a small number of more basic and fundamental axioms?

3. *The Scale of Consistency.* Are the various elements of the worldview
   consistent with each other?
4. *The Scale of Plausibility.* Is the worldview, in its concepts, theories,
   and values consistent with the best available facts, theories, and models
   of contemporary science?

Using these four criteria, we can trace the emergence of environmentalists' varied worldviews and, in the process, deepen our understanding of
the perceptions of environmentalists by better understanding the ways their
vocabularies mediate between perceptions and actions.

EXPLOITATIONISM

Muir and Pinchot did not act in a cultural or political vacuum. The social
context in which they acted was largely determined by a dominant worldview, which might be called "exploitationist." A special though not unique
set of conditions appeared when a frontier was opened to the relatively
populous nations of Europe by the colonization of the New World.[3] The
atmosphere of opportunity and rising expectations that accompanied the
new availability of land encouraged a careless attitude about resources among
the early colonists.[4]

From early colonization until middle of the nineteenth century, wilderness areas and raw natural resources were therefore seen as uncontrolled
by man, unproductive, and valueless until human labor was mixed with
them.[5] Transforming trees into lumber, wilderness into tillable land, and
metal ores into tools were all viewed, without qualification, as good acts.
They made the useless useful. Roderick Nash, in discussing the attitudes
of early colonists, quotes one who appealed to the idea of man's God-
appointed dominion in concluding that he did not see "how men should
make benefit of [vacant land] . . . but by habitation and culture."[6]

In this context a worldview that can be labeled "exploitationism" prevailed—all scarcity was scarcity of humanly useful goods, tools, and services. Exploitationists perceive no shortages of raw natural resources because value was imparted to these resources only by the addition of the
truly scarce resource, human labor. It was therefore conceptually impossible to speak of "waste" as a by-product of transforming raw resources.

This is not to say, of course, that early exploitationists had no concern
for the future. Indeed, commentators have made much of the vision that
was shared by early colonists, referring to their perception that they were
engaged in an "errand into the wilderness"[7] and their resolve to "build a
city on a hill." Early colonists were often critical of waste and of sinful
materialism.[8] These criticisms, however, were directed at conspicuous con-

sumption and at an unseemly emphasis on material comforts. Concern for the future emphasized the need to create durable goods and capital that would be useful to future generations in the pursuit of prosperity—but this concern did not extend to resources.

The colonists' attitude of abundance and its concomitant emphasis on the value of labor also placed high value on large families and encouraged population growth. The changing patterns of American attitudes toward population growth were therefore intimately intertwined with attitudes toward the value of resources. Further, both of these attitudes were deeply affected by frontier conditions—as long as Americans saw a limitless frontier of land, worries about overpopulation were inconceivable.

The colonists' concerns were practical, but it may be helpful to characterize their exploitationist worldview as embodying two axioms. I will call them the "Axiom of Usefulness" and the "Axiom of Abundance." They are axioms in the sense explained above: They are unquestioned principles that underlie and shape the other elements of the worldview. The Axiom of Usefulness asserts that production of goods for human use is a good thing. It supports exploitationism only when coupled with the Axiom of Abundance, which implies an assessment of the comparative role of raw resources and labor in production—any natural resource, prior to its transformation by humans into a "product," can be replaced by a substitute resource, without significantly increasing costs of production. In other words, shortages of raw materials are not an important factor in determining the value of products.[9]

The first axiom embodies the value of productivity for human use and the second formalizes the attitude that waste of raw resources involves no true waste at all. When working in tandem, these two axioms create an attitude toward growth and development that can support no moral disapprobation of "waste" of raw products or the systems that produce such products.

CONSERVATIONISM

The attitudes that set the stage for a conservationist reaction in America beginning in the second half of the nineteenth century were thus intimately tied to the closing of the American frontier—they represented a reaction against the dominant attitude of exploitationism.

Environmental protectionism, in its beginnings, must therefore be understood as a reaction against the dominant social trends of privatization and exploitation. While it is difficult to imagine two people more different in temperament and outlook, Muir and Pinchot both reacted at first to a shared perception that, unless farsighted citizens took up the cause, private interests aided by compliant government agencies and policies would leave the nation denuded, eroded, and aesthetically degraded in pursuit of individual profit. In these early, shared concerns, conservationists were mainly *reactive*. They responded to immense destruction with disapprobation—the common denominator of their reaction was moral disgust, although they chose quite different vocabularies to express that disgust.

Pinchot expressed his reaction in the language of scientific resource management. He recognized that the exploitationist attitude, which had held the whole country in its grip during the frontier era, was inappropriate for a more mature nation that could no longer feed growth by devouring wilderness and untapped resource potentials at its frontiers. His conservationist worldview, if it is analyzed more theoretically than Pinchot was inclined to do, can be characterized as rejecting the Axiom of Abundance but never questioning the Axiom of Usefulness. Pinchot believed that waste would imminently lead to shortages of essential resources, such as timber, minerals, and productive land, and that vigilant replacement of renewable resources, such as timber, was necessary. The language of economics therefore became the language in which he expressed his new worldview; that language and its reductionistic vocabulary shaped and gave cohesion to it.

Pinchot's worldview was reasonably well connected, resting as it did on the language of economics as the sole expressor of management goals; that language linked his utilitarian value system—the Axiom of Usefulness, which he interpreted democratically and progressively—to those management goals. This was the utilitarian insight that Pinchot "discovered" while riding his horse Jim in Rock Creek Park.

But Pinchot's worldview, while well connected, was radically incomplete. The language of economics extant in Pinchot's day focused much more brightly on materialistic values and commodity production; we noted that it was hard for him to see, much less to emphasize, broader aesthetic values such as stream protection. Remember, Pinchot refused even to refer to New York's preservation of the Adirondack Forest as "management," because it produced no timber. Management, to him, was economic management, and economic management was commodity production.

When Pinchot *defined* resource conservation as maximizing the material well-being of all the people, he illustrated how largely unexamined semantic assumptions embodied in worldviews determine the application of important values. By assuming the Axiom of Usefulness and reducing the task of resource management to one of regulating and protecting the flow of natural resources, he naturally concentrated his attention on supplying the raw resources necessary to fulfill economic demands. Because material productivity was an unquestionable good for Pinchot, he saw scientific resource management as means-oriented and value-free. The language he adopted was a means-oriented language of economics.

Politically, however, he had also to appeal to the apparently independent principle of fairness in distribution. Pinchot's personal touch was to give the attack on the Axiom of Abundance a populist, democratic flavor. He emphasized the protection of resources for all the people, present and future, and—not unrealistically, given the economic and political situation— he saw his task as protecting resources from the unrestrained greed of a few wealthy exploiters. He therefore emphasized the importance of retaining forest reserves in public ownership.

This grafting of populist egalitarianism onto economic reductionism avoided a crucial question: Might population growth, even of economic equals, lead eventually to exhaustion of resources? Pinchot and his early followers never developed a position on the limits of equitable exploitation—it was left to Leopold to apply the concept of a "carrying capacity" to human use of the land.

MUIR AND PRESERVATION

For all his rhetorical brilliance, Muir never succeeded in constructing a well-integrated worldview to place in opposition to that of the exploitationists or to that of Pinchot. Muir's worldview, like Pinchot's, had no room for the Axiom of Abundance. But Muir's response to the Axiom of Usefulness was far more complex than Pinchot's. He never questioned that nature was useful; Muir was, after all, an accomplished machinist and orchard operator. He objected, however, to the usual implication that nature was valuable *only* because of its human uses—he saw this anthropocentrism as arrogant, egotistical, and insensitive to the needs of other creatures. While he appealed, in an early unpublished manuscript, to the rights of rattlesnakes and plants, in his published writings Muir justified the preservation of the beauties of nature because he thought experience of nature cures the alienation of modern society. He saw humans as capable of a consciousness higher than the make-a-buck mentality of the exploitationists, who emphasized only production and consumption, and was convinced that communion with the natural world and its beauties would promote that higher consciousness.

Muir's attempts to "save" the Sierras were therefore given two apparently distinct justifications. Muir thought the Sierras needed to be saved from human greed in the form of profit-motivated developers and concessionaires. In this mode, Muir spoke of saving the mountains for themselves, quite independently of human motives. He rejected the Axiom of Usefulness because of its human-centered attitude. But the second goal of the Sierra Club, mentioned in the original charter proposed by Muir in 1892, was to increase access for people to the beauties of the mountains. When emphasizing this latter goal, Muir saw wild nature as spiritually instrumental, as a means to inspire awe and to explode the exploitationists worldview. His view of river valleys as holy places implies a less general modification of the Axiom of Usefulness—nature can be preserved by recognizing nonmaterial and nonconsumptive human values, such as aesthetic enjoyment and spiritual fulfillment. In this latter mode, Muir followed Thoreau and Emerson in arguing that modern industrial society caused alienation and that experience of nature healed the ills besetting moderns. Muir's worldview, therefore, can be characterized as definitely rejecting the Axiom of Abundance and as also attacking the Axiom of Usefulness as inadequate for two reasons: It ignored human spirituality *and* was based on an assumption, "totally unsupported by the facts," of anthropocentrism.

We can, in cases like this one, say that the policy positions of environmentalists are "overdetermined;" more than one value axiom is sufficient to support a single policy position, in this case, preservationism. While there is no obvious inconsistency in this case—wild nature may have value of all these types—Muir's employment of these varied fragments suffered because he never *integrated* these two disparate arguments with each other, or with his grudging acceptance of utilitarianism, into an overarching, logically reconciled system. We saw, in Chapter 2, that Muir felt all of his values would be unified by pantheism, which he took to imply that all species have value in the great scheme of things, even though he never provided precise rules for balancing human interests with the interests of other species. In cases like this, we can say that environmentalists' worldviews do not do very well on the second scale, the Scale of Connectedness. Since the elements of their worldview are not connected to each other by central, organizing principles, they appear to apply varied worldviews and values willy-nilly.

Lacking a decisive principle of integration for his three apparently independent value axioms, Muir applied a variety of principles to guide land use in an ad hoc manner. These criteria, one based on enlightened utilitarianism, one based on human spiritual needs, and a third based on extrahuman values in nature, were never fully connected in Muir's worldview. This lack of connectedness, together with a tendency toward isolationism, prevented Muir from providing a plan for integrating human uses of land into the larger landscape. He was clear that some areas such as Yosemite, at least, should be sacred places. But he provided no criterion for separating sacred places from productive ones and, since the sacred places would be used for human aesthetic and spiritual enjoyment, he never faced the problem, so serious today, of how to provide outdoor recreation to an expanding population without destroying the very natural characteristics that it depends upon.

Whereas Pinchot's professionally determined worldview and the economic language in which it was expressed placed blinders on its users, making them unable to see the importance of spiritual and aesthetic uses of nature, Muir's worldview, by contrast, was extremely rich. Acting as an amateur, Muir never developed a specialized professional worldview. Muir's worldview afforded a rich rhetoric, but a rhetoric torn by internal tensions: For example, Muir really did not know what to say about the limits of the wise use of resources. He clearly recognized that we must exploit to live, and at first he praised Pinchot's wise-use management. He clearly thought that his pantheistic theology provided a general metaphysical and moral system capable of comprehending and integrating all of the values he saw in nature. But in the end, he chose a moral rhetoric, pitting the forces of good against the forces of evil, which conceived of developmental plans as sins against a higher law. That rhetoric left no room for integrating legitimate uses of nature, since it implied that humans harm nature in all their manipulations and exploitations. Muir's moral rhetoric and his demand for "righteous management" left him no basis for a distinction between acceptable and unacceptable exploitation.

One might say that Muir's worldview scored not so well on the Scale of Consistency, but here we must be very careful not to confuse two related issues. Muir cannot be accused of direct inconsistencies—his religious rhetoric was not precise enough to create direct contradictions. It is more accurate, therefore, to say that the internal tensions in Muir's worldview represent a lack of connectedness among his ideas. He had at least three independent axioms of value—some not very clearly articulated recognition of the usefulness of nature to humans, a commitment to the "spiritual usefulness" of nature to humans, and a belief that nature, viewed in the largest perspective, was God. The latter two elements were not integrated with the former one, but they led him to abandon any pretense of value-free "science" and to employ a religious rhetoric that did not mesh smoothly with any use of nature at all.

TOWARD INTEGRATION

Environmentalists today have rejected Pinchot's simple reductionism and aggregationism, opting to follow Muir's approach of embracing multiple value axioms. The environmentalists' dilemma and its divisive effects are the result. Some environmentalists, such as academics who urge "biocentric egalitarianism," prompt activists to narrow their value foundations and adopt a single value system to the exclusion of others. These academics seek a worldview with greater connectedness. But, among active environmentalists, political pressures to build coalitions encourage them to maintain the broader appeal supported by a variety of value axioms. The necessities of political rhetoric have usually overcome the niceties of logic.

To see how differing moral principles and visions have functioned in the modern era, it is instructive to compare Aldo Leopold's career with that of Rachel Carson; these two careers can be thought of as mirror images.[10] Leopold began as a forester involved in day-to-day policy discussions. In spite of an early and lasting personal attraction to nonanthropocentric ideas, he generally played these down in his interactions with policymakers and government officials. Leopold justified this approach by embracing a tolerant philosophical attitude and by recognizing that human-related reasons are likely to have a more immediate effect in policy circles. Toward the end of his career, he placed more and more emphasis on the need to embrace a new and more appropriate ethic—a land ethic—and the nature essays in *A Sand County Almanac* represented his attempt to alter posterity's conceptions and perceptions of the world. He explicitly recognized that the adoption of a land ethic would represent a major shift in human consciousness.

Carson began where Leopold ended up, as a nature writer; her early efforts were intended to alter the perceptions of her readers. Mostly, she employed brilliant description, but she also employed ecological ideas to challenge human arrogance toward other species. Faced with a threat that the natural world she so loved would be overwhelmed, "silenced," by careless use of pesticides, she was driven into the arena of concrete public policy discussion, reversing Leopold's path. Once she determined that she must attack the misuse of pesticides, she used ecological arguments to emphasize the threat of chemicals to human health and survival. She injected ecological concepts into the common vernacular, warning that humans, who eat high on the food pyramid, are especially susceptible to persistent chemicals that can bioaccumulate as they work their way up the food chain. Leopold and Carson no doubt did more than anyone else to shape the thinking of modern environmentalists, and they shared a willingness to tailor their arguments to relevant constituencies. Both were committed to a biocentric perspective and attacked human arrogance. When they found themselves in political and policy arenas, however, they emphasized human-oriented arguments. And they did so without apparent remorse and with great effect on events.

I am not, of course, accusing Leopold and Carson of insincerity. Not even their most vocal critics would have accused them of that—overzealousness, perhaps, but never insincerity. No, their easy transition between worldviews requires an ideological explanation. If we can understand the logic of this transition, it may help us to understand the value pluralism of the environmental movement more generally. If we can understand how Leopold and Carson could embrace biocentrism, and yet argue in public from explicitly anthropocentric principles, it may help us to understand how environmentalists who embrace highly divergent ultimate values and apparently differing worldviews can coexist in an influential social movement.

When Leopold and Carson shift back and forth between anthropocentric utilitarianism and biocentrism, for example, their shifts cannot be understood within closed—that is, complete and highly connected—systems of thought. We saw, in the last chapter, the empirical dangers of assuming that environmentalists operate in reality with closed, opposed systems of thought. Here, we see the logical weakness of closed paradigmatic analyses. Those analyses cannot illuminate the transitional moves Carson and Leopold made *between two systems of thought*, because choices of which value system to emphasize in a given situation exist, by definition, *outside* either system.[11]

It is now possible to recognize an important ambiguity in the Scale of Connectedness which, loosely stated, places a value on having a small number of axioms that unify and give internal structure to a worldview. Our loose, preliminary statement of this criterion for judging worldviews might be taken to imply that an ideal worldview—one that is maximally connected—would have only a single moral axiom from which every value judgment must ultimately derive. While a system based on a single principle is unquestionably elegant, and furthermore minimizes the intrasystematic tensions that threaten contradictions, the goal of connectedness must be kept in balance with the criterion of completeness. Pinchot achieved connectedness in his worldview by reducing all human values to utilities associated with democratically distributed material consumption, but at the cost of blinding himself to aesthetic and other important values. In embracing Muir's rich, multiprincipled worldview, modern environmentalists saddle themselves with internal tensions, but they gain a rich vocabulary capable of expressing many values and appealing to many constituencies.

Nor is it obvious that environmentalists, by so doing, must doom themselves to incoherence and self-contradiction. Competing values can, presumably, be reconciled within a single worldview if there exist integrative principles that determine which values should take priority in any given situation. Leopold, I have asserted, at least began the process of conceptualizing the necessary rules of integration: productivity values have free play until their pursuit threatens the larger context, at which point limits, to be articulated in the ecological terms of system fragility, constrain choices based on a pure productivity criterion.

But Leopold's integrative theory was never made precise, and ecological concepts may not prove capable of articulating integrative principles adequate to balance, in all situations, the multiple values expressed in the varied languages and worldviews of environmentalists. Leopold's approach nonetheless represents a good start. Until it is demonstrated that Leopold's prototypical integrative system cannot be consistently applied, environmentalists can defend their shifting use of varied rhetorics and value principles as characteristic of a rich and complete worldview. They eschew Pinchot's elegant single-principled approach and avoid the aesthetic and other blind spots entailed by his reductionism.

Biocentric egalitarianism, on the other hand, has proved politically nonviable, unless it is supplemented in many contexts with human-oriented rhetoric. A rhetoric that ignores human interests does not appeal to a broad enough slice of the American public to support a politically effective movement. Environmentalists therefore appeal to a variety of value systems and languages, emphasizing one or the other in differing situations. Meanwhile, rhetorical confusion hampers the attempt to articulate integrative principles. Two points are clear. Those integrative principles, once articulated, will be stated in an ecological vocabulary, and they will rest on moral as well as purely economic values. Beyond these two points of agreement, confusion reigns.

But there has been another important by-product of the Leopoldian consensus. Conceiving management problems as problems of ecological context has forced environmentalists to adhere to relatively strict standards of scientific objectivity, even as they argue for the mandates of their management tasks in moral terms. Regardless of questions of rhetorical style, Leopold and Carson set high standards for scientific objectivity, standards environmentalists maintain today. Because their arguments, including Leopold's argument against atomistic management and Carson's argument from bioaccumulation, required premises of considerable generality, activists were forced to expose themselves to considerable risk that their premises might be undermined by further scientific data. While there are instances in which environmentalists have too quickly endorsed scientific generalizations,[12] on the whole their arguments have scored high on the plausibility scale. Furthermore, while they outrun available data in some cases, they have been quick to correct any formulations that become scientifically questionable, and they have amassed an impressive number of victories in administrative hearings and court cases that turned on complex scientific arguments.[13]

To sum up, environmentalists have gained considerable clarity and consensus in some aspects of their worldviews, especially regarding the importance of ecological science in management; but even today they employ a variety of rhetorics that can be traced to the multiplicity of ultimate value axioms originally encompassed in early protectionists' varied reactions against extreme exploitationism. They appeal, unsystematically, to a variety of value axioms in different situations. Leopold's contextual integration, while unquestionably bold, put environmentalists in the position of needing both staggering amounts of scientific data and sophisticated theories in order to manage intelligently. They placed a huge burden on the struggling discipline of ecology, a burden it was often unable to bear. In particular, the problem of parts and wholes—what unit to consider a "whole" ecosystem—often left environmentalists perplexed on the very questions of scale that, according to Leopold's contextual system, held the key to good management. And since questions of scale are so intimately tied to management goals, values guiding management decisions cannot be separated from choices regarding the ecological unit to choose as the managerial unit. This conclusion represents one more way of saying that conservation biology cannot be a value-free discipline. It also emphasizes the tremendous task environmentalists face in developing a complete and integrated theory of environmental management.

The weakness of ecological concepts sometimes encourages environmentalists, again in the tradition of Muir, to employ mystical, theological, and metaphysical language to fill the gaps in ecological theory. In these areas, environmentalists have not progressed much past Leopold's brilliant 1923 analysis of the prospects of developing a unified worldview to guide management, an analysis that was expressed mainly in metaphorical terms. Leopold concluded that establishing one of the various conceptions of the world as certainly true stretched the limits of our language, but that wise management, to protect the ecosystematic context of all human activities, follows from a variety of "conceptions" including organicism and noble anthropocentrism. Even today, environmentalists are more confident regarding what should be done in specific situations than they are in choosing a vocabulary and attendant worldview with which to justify those policies; they lack confidence in the latter choices because they lack precise rules of integration for determining the relevant managerial context and the appropriate managerial goals in particular situations.

Leopold's contribution was to propose, in very general terms an integrated, multilevel synthesis that was pluralistic but emphasized different values in different situations, depending on the impact of economic activity on a broader, ecological system. Leopold adopted an ecological vocabulary for discussing environmental problems, subordinating Pinchot's resource management approach within a larger worldview reminiscent of Muir's pantheism. But the ecological language held out the hope, however distant because of the complexity of ecological systems and their recalcitrance to theoretical mastery, of an art of management based on a holistic science that comprehended human economic impacts as impacts on the larger ecosystematic context. This integrated approach, while still imprecise, avoided direct policy conflicts in many actual situations. Provided environmentalists could agree in practice whether to emphasize wise use and the values of sustainable production or broader ecological criteria in a given situation, policy positions could be justified according to a variety of worldview fragments, including biocentrism, pantheism, or enlightened utilitarianism.

Philosophy's Road: A Methodological Interlude

For my purposes—to explore the hypothesis that environmentalists are evolving toward a consensus in policy, even though they remain divided regarding basic values—requires a special philosophical method. It will not do simply to create abstractly defined systems of moral reasoning. If those values play no role in activists' reasoning, there will be no illumination of how environmentalists' values interact so as to support an emerging consensus. Nor can I pursue a *descriptive* method that would simply record the values environmentalists articulate—those values may be incomplete, confused, or not articulated at all explicitly. I must of course show, by reference to environmentalists' actual assertions, that they appeal to differing values. And I must show, by reference to policy positions advocated, that they are indeed moving toward a policy consensus. Merely to demonstrate these points, however, would be unsatisfying, and here is where philosophical analysis of an uncommon type becomes necessary. We want to know not just what justifications *could* be given for a policy or just what values are in fact held: We want to know *how* and *why* individuals with importantly different values are gravitating toward similar policy positions. If we are to *understand*, not just record, the policy consensus that is emerging against a backdrop of value pluralism, we must explore the interanimation of values and policies in actual situations. We must understand the variety of values environmentalists promote as parts of larger worldviews, which are related to action and policy on the one hand, but on the other, also embody broader metaphysical and moral assumptions that provide the context of management decisions. This larger arena of worldview analysis may provide clues not only to the values environmentalists hold, but also to the second-order principles that guide decisions as to which values should be given priority in which situations.

For purposes of comparison with the uncommon sort of method we need, I will briefly present an example, in the form of a teaching device I have found successful, of the usual approach of philosophers. Whenever I teach a course in practical ethics and need to introduce students to basic methods of ethical reasoning, I open class discussion with a hypothetical situation.[14] Projecting the students back into frontier days, I ask each of them to imagine that he or she is Commandant of a fortress deep in Indian territory, extremely remote from sources of reinforcements. After years of friendly relations with the Indian tribes in the region, the fortress has defenses that have deteriorated. The cannons are old and unreliable; ammunition, mostly used for target practice and an occasional hunting trip, is in short supply. But the student is asked to imagine that, upon return from a three-day trip with the second-in-command (we call him Jones), the Commandant finds the fortress surrounded by Indians in war paint and formidable Indian armaments. Once inside the fort—the Indians part to allow uncontested passage—the Commandment pieces together what has happened. Subsequent to Jones's departure with the Commandant, the Chief-of-Chief's eldest son was foully murdered. The Chief consulted with the head Seer of his tribe who, in turn, consulted the entrails of a freshly slain (prairie) chicken. The verdict, the Seer concluded, was incontrovertible; Jones had killed the Indian Prince.

The Commandant, knowing Jones was with him, far from the crime-site, is certain of Jones's innocence. But the Chief brushes the alibi aside hastily; the Seer's methods cannot fail. The verdict has been rendered and with it an ultimatum. Either Jones is turned over by nightfall for Indian justice (to be skinned alive) or the unified tribes of the Indian Nation will attack the near-defenseless fortress. All white men and women who are not killed in battle will, like Jones, taste Indian justice. Each student, as Commandant, must decide what to do and explain why.

Discussions of the hypothetical case invariably proceed in three stages. In the first stage, students propose ways out of the quandary (and I reject these). "Send for reinforcements" (there isn't time); "reason with the Chief" (he's finished talking and he is counting the hours until sundown); "ask Jones if he won't, out of community spirit, give himself up" (Jones, busy cleaning his rifle, says, "No way"); "volunteer to go in Jones's place" (the Chief will hear none of it—it's Jones or everyone). This first stage, in other words, always involves a series of challenges to the assumed facts that constitute the quandary. The students want, quite reasonably, to find a way to avoid the excruciating either-or decision.

In the second stage, students choose sides: give Jones up or fight for what is right. When I have asked for a show of hands, surprisingly, there's usually a fairly even split. Surprisingly, because as the example was initially explained to me, it was to be a knock-down case where an innocent individual should be sacrificed for utilitarian reasons. (Jones of course suffers the same fate either way—it should make little difference to him—and presumably the hundreds of other occupants of the fort have strong preferences for Jones to experience Indian justice alone). Nevertheless, I usually find at least as many students who opt to fight.

Finally, students begin noting that two very different styles of reasoning seem to favor one choice over the other. Some students even mention the class readings. If we, with Jeremy Bentham, emphasize the *consequences* for individuals, computing the utilities of the two alternative courses of action, we tend to say, "Sorry Jones, but . . . ." If, on the other hand, with Immanuel Kant, we emphasize that moral choice is a matter of discharging one's obligations (the view of ethics often called "deontology"), it seems that the Commandant must uphold our cherished moral principles, to protect the innocent and uphold the rule of law.

At any rate, students seem to prefer the discussion to a lecture on Kant and Bentham, and their occasional references to duties and to utilitarian calculi encourage me (shamelessly) to use the example over and over, despite its unfortunate suggestion that Native American decision processes were arbitrary and capricious. Here, I use it once again, but for a different purpose. I want to discuss *why* the example works and what it tells us about the usual methods of philosophers. The example is, admittedly, cooked up. It achieves its pedagogical purpose *precisely because it is cooked up*. If it were a real case, we would never get beyond the first stage of discussion—because both alternatives are so unattractive, nobody in a real situation would give up on somehow subverting the dilemma. But the defining factual conditions of the case—the remoteness of the fortress, the hopelessness of the fight, the intransigence of the Chief—are designed precisely to force students' reasoning into one of two classical modes of moral reasoning. They are driven back upon their moral resources (either their intuitions or a cursory reading of Kant and Bentham) by my dogmatic insistence on just these facts. The students must make an unavoidable choice between two stark options. Therefore, by fixing the facts as extreme and indisputable, the example focuses attention on two quite different methods of justification for moral choices.

Although the subject matter is different, the purpose of the class discussion resembles the goals of philosophers such as Naess and Passmore, who we found in the last chapter stating opposed, highly connected moral positions for the purpose of evaluating the strength of various styles of argument for environmental policy positions. Naess and Passmore have in mind examples in which, by telling a fairly detailed story, environmentalists are apparently forced to choose between the interests of human beings taken as a whole and some element of nature. One such example that is sometimes used is the example of the snaildarter, a tiny three-inch fish whose only known habitat was to be swamped by the Tellico Dam. If one assumes that the dam is unquestionably a good thing for humans and that the dam will cause the extinction of a species of fish, then we are forced into an excruciating either-or decision analogous to the fortress example. In both cases, the purpose is to evaluate the reasoning processes by which one gets from certain carefully defined moral principles to policy proposals. By pretending, contrary to fact, that environmentalists operate with well worked-out, essentially closed systems of moral principles, and by dealing only in carefully confined cases in which cooked-up facts ensure that human interests conflict with those of other species, it is possible to focus attention on the argumentative strength of the moral axioms that constitute the system. Sharply isolated principles and artificially constrained cases allow us to trace a more direct road from moral principles to actions.

But it is useful to notice that the artificially constrained situations are not at all like normal situations in which we face moral choices. Seldom are the facts unquestionable, seldom does a situation face us with a forced choice between only two options, and seldom do moral principles such as utilitarianism and deontological reasoning (or anthropocentrism and biocentrism) point in directly opposing directions. In actual situations we usually can cite conflicting data, we can choose from an array of options (including inaction), and, more often than not, competing principles suggest similar moral strategies. Therefore, it is not surprising that I must put on my most imperious teacher-attitude to force students to take the example at its face value. If there were a chance to contact reinforcements, if the Chief would negotiate, if virtually any other options were open to the Commandant, then both utilitarianism *and* deontology would imply that we choose those options over either fighting or surrendering Jones.

We will see, in Chapter 8, that the snaildarter case, though unquestionably a real case and unquestionably useful for the purposes of clarifying environmental reasoning, does not, in actuality, exemplify a choice between human interests and the existence of a species. First, it turned out that another large and healthy population of snaildarters exists in another branch of the river unaffected by the dam. Further, once the snaildarter case became a cause célèbre, a high-level committee insisted on a new benefit-cost analysis of the dam project. The General Accounting Office completed the new analysis and found that, quite independent of the threat to the snaildarter, the dam would be a losing proposition—it was a pork-barrel project supported initially by a trumped-up benefit-cost analysis that greatly overestimated the human benefits and greatly underestimated the human costs of the project. Thus, in order to use the snaildarter case as an example of a stark choice between human and nonhuman interests, one must hedge the case all around by qualifiers: Suppose there is a case just like the snaildarter case, *except* there are indeed no other snaildarters in the world and *except* that a legitimate accounting of human interests does in fact favor the dam, and so on. That is, the snaildarter case provides an example of human interests in conflict with another species' interests only if one adds contrary-to-fact assumptions to circumscribe the conditions of the case.

Now, given the purposes of this book, to examine why environmentalists who espouse diverging values nevertheless support converging policies, we must pay careful attention to actual policy situations because, it will turn out, in actual situations there are many cases in which anthropocentric utilitarianism and biocentric egalitarianism imply the same policies, just as utilitarianism and deontology will often point to the same choices in real-life situations. Thankfully, few of our decisions are as stark as the one faced by the Commandant.

In the next part, four chapters will be devoted to searching for policy consensuses in four general areas of environmental concern: resource use, pollution control, protection of biological diversity, and land use. My purpose will not be to ask whether differing worldviews and different basic value systems that environmentalists *might* adopt (if they were more systematic thinkers than they are) would lead them to adopt the same or similar policies in some constrained, artificial situation. My purpose, rather, will be to examine the actual conditions under which environmentalists advocate similar policies and justify them by appeal to different basic values.

The method can be explained metaphorically—philosophy is the search for a road, a road that connects metaphysical conceptions of the world and basic moral axioms with actions. Often, when philosophers' purposes are to delineate axioms and evaluate their argumentative power (as was my purpose in the fortress example), they build fences right alongside the road, limiting the factual uncertainties and forcing reasoning into narrowly confined paths. They also emphasize going in only one direction on the philosophers' road: from sharply delineated values toward actions. But in fact philosophy is not so self-sufficient as this unidirectional method would suggest.[15]

John Rawls, in his influential *A Theory of Justice*, describes the philosophical method he uses in searching for a "reflective equilibrium" between principles of justice and our most firmly held intuitive convictions about what we should do in particular situations.[16] He describes the method as, first, defining a set of moral principles that represent a first try at a theory of justice, conceiving these as principles that "rational persons concerned to advance their interests would consent to as equals." Then we must "see if the principles which would be chosen match our considered convictions of justice or extend them in an acceptable way."[17] That is, we check the principles against our moral intuitions concerning what would be right or wrong in a particular situation. If there are discrepancies between the implications of our principles and our firmest convictions of what is right or wrong in a particular case, we adjust either the principles or our intuitions until they square with each other. We have then arrived at a reflective equilibrium. The equilibrium is unstable, however, because further reflection may cause us to adjust our principles, or a new case may be suggested in which our principles clash with our intuitions.[18] If this happens, then we readjust, seeking a new equilibrium.[19]

Because, as Rawls notes, principles "cannot be deduced from self-evident premises or conditions on principles," this method is the best available. We cannot first state and fully justify our principles and then apply them.[20] According to Rawls's method, therefore, philosophy's road is not a one-way street—we can learn more about and correct the principles we embrace by paying attention to particular cases, as well as systematizing our intuitions about cases by developing our principles. Perhaps we can conceptualize our method, and philosophy's road, as a large circular drive. We can start at principles, pass through case-applications, and return to principles; or, we can start at cases, pass through principles, and return to cases. With luck, successive trips around the circle will result not in dizziness, but in an upward spiral of understanding.

The method used here will resemble Rawls's method in general outlines. Two qualifications are in order, however, due to the particularities of our inquiry. First, because this is a book in practical ethics, concerned with how principles are applied in the real world, we will emphasize *actual* cases, rather than the carefully constrained hypothetical cases, such as the fortress example, often employed by philosophers when they are concerned mainly to test and refine principles. Since environmental activists usually operate with open systems, we cannot expect all examples to fit a neat and simple pattern.

Second, because I am hypothesizing that, in the case of environmental values and environmental actions, there is greater agreement about what to do than there is about which principles justify it, we will have to be (at least initially) tolerant of appeals to a variety of principles. At this point, we face a choice. We could assume that the disparity of values represents an immaturity in the search for principles, and that once we get the principles right, there will emerge a single set of principles that all true environmentalists will accept. It seems that this assumption motivates a number of environmental ethicists, such as deep ecologists, to be highly critical of environmental activists who ascribe to principles other than the ones the philosophers favor. Indeed, this assumption has motivated much of the writing in environmental ethics in the first fifteen years of its existence as a discernible field. Naess, for example, sees "shallow ecology" as embracing unacceptably anthropocentric values, while Passmore argues that non-anthropocentrists are committed to an incoherent theory of value.[21] Of course we can learn much from analyses that assume there is only one ultimately rational circle connecting values and actions that environmentalists can traverse.

But in this book we will not make this assumption, leaving open instead the possibility that environmentalists may, indefinitely and without irrationality, go around more than one circle.[22] It will be interesting if, in traveling different circles through different sets of principles, environmentalists nevertheless arrive at the same, or nearly the same, points on policy. If that is the case, as I am hypothesizing, it will be interesting to ask why. It may be that, in the particular area of practical ethics known as environmental policy, there are special features of the terrain that force traffic around several different circles to intersect at the same policy points. If so, we should pay special attention to those features; in the process, we may learn something about ethical principles as well as about environmental policy.

By thus going from practice to theory, philosophy can contribute to the search for an integrated worldview, a common language, and a more adequate shared conceptualization of the world. By first describing an emerging policy consensus in Part II and, only then, returning to philosophical principles in the final part, value analysis is trained specifically upon the goals and values environmentalists in fact espouse. The goal of the empirical description is to look for common-denominator objectives—goals that virtually all environmentalists support. Then we will codify and systematize these objectives in various areas of environmental policy, showing how these objectives can be expressed, explained, and justified by appealing to a set of philosophical axioms—the elements of a worldview (the system of axioms that gives shape and meaning to environmentalists' perception) that would support these consensus objectives. This is a method of searching for a new, shared worldview. Because it goes the other way around the action-theory circle, this study is designed to find common ground, and to go as far as possible toward consensus.

Environmentalists' emerging consensus, it will turn out, is based more on scientific principles than on shared metaphysical and moral axioms. Science, Leopold concluded, supports the need to protect the ecological context. If we fail to do this, we will destroy the context of our cultural values. Whether one sees nature as created or evolved is not so important in policy as the *shared* goal of protecting the context in which all values are given meaning. The emerging consensus in environmental policy can therefore be based on a variety of philosophies. One feature of this approach is that it can be pluralistic, even while seeking to develop a shared worldview—it may turn out that environmentalists have considerable freedom of intellectual choice regarding the philosophical aspects of a worldview. Leopold, remember, argued that the search for a land ethic is a creative cultural adventure and advocated management principles that could be explained on a number of alternative "conceptions" of the world.

If we are to explore these special features of environmental policy, we cannot simply employ the rational, evaluative methods of the philosopher who proposes a single set of principles and tests them against carefully constrained artificial cases designed to focus attention on specific outcomes of carefully stated principles. Nor can we simply *report* what policy positions environmentalists support and how they justify these in moral terms. We must rather pay attention to the policies environmentalists espouse, try sensitively to relate these policies to the moral principles by which they justify them, *and* examine very carefully the situations in which environmental policy is formulated. If my central hypothesis is true—that a policy consensus is emerging amidst moral pluralism—we should find special features of today's environmental situation that explain why environmentalists who pass through differing ethical theories nevertheless travel on roads that intersect at the same policy points. If we hope to find this explanation, we will have to examine environmental action, environmental discourse, and all of the points at which values affect both action and discourse. A model sufficiently inclusive to focus on this entire milieu will be presented in the next section.

Closed and Open Systems: An Action-Oriented Approach

We have referred to "worldviews" of environmentalists rather than following social scientists in referring to "paradigms." While both of these concepts assume that our beliefs and values hang together as interconnected constellations, the concept of a paradigm is also associated with another, stronger assumption—that environmentalists accept or reject one of two competing paradigms monolithically. That assumption has led to confusion in empirical work, especially to a tendency to reify exclusive and opposed categories of environmentalists, when environmentalists themselves have emphasized their policy agreements, and have generally been inclusive and pluralistic in their approach to values.

This stronger assumption, which we saw operating in the work of Naess, Devall, and Milbrath, can be understood as implying that the worldviews of environmentalists are "closed," in the sense that they are consistent, complete, and highly connected systems of principles. The key difference between a worldview and a paradigm is that worldviews are thought of primarily as guides to action, whereas paradigms are systems of thought— concepts, beliefs, and values. A worldview analysis is, therefore, more inclusive than a paradigmatic analysis in that it must ultimately take into account the varied roles—explanation, justification, advocacy, apology, and so forth—that values and beliefs play in guiding environmentalists' actions. Philosophers are usually content to discuss the consistency, completeness, and connectedness of systems of thought that guide environmentalists; but we saw, in the last section, that this method, while useful in evaluating the quality of argumentation, cannot serve our exercise in practical philosophy. Philosophy's road, in such exercises, must pass through discussions of concrete, real-world cases and policy debates, for it is there that we can really see the varied roles that values play in the worldviews *and actions* of environmentalists.

What is needed is a pluralistic model for understanding how environmentalists' worldviews affect action. For this purpose, I suggest that we construct a model, a simplified structure, for understanding environmental action. A model in this sense is more inclusive than a paradigm or a worldview, both of which emphasize clusters of values, attitudes, and beliefs. A model examines how worldviews, which are understood as guides to actions, function as part of a larger structure, which includes also an action component and a political-interactive component.

Such a model must take into account the reactive nature of much environmental activism. To ensure this, I will introduce the open, pluralistic model with a simplified hypothetical example.[23] Imagine a rural community that loves and takes immense pride in a designated recreational area that represents the character and spirit of the community. While the recreation area includes a number of township, county, and state parks, it also depends for its character on surrounding, large blocks of privately owned forests. Now, our case begins when the largest private owner of forest lands, an out-of-state holding company, buys and upgrades an old sawmill and declares its intention to clear-cut its huge forest holdings for timber. This development is especially upsetting to the community because the headwaters of the river that is a focal point of several parks and campgrounds are on the holding company's land. By looking at a series of events that might occur in response to such a situation, we will be able to isolate several stages of environmental action—these stages will, in turn, represent the main elements of our model.

The first stage of our model represents the *perception of a grievance*. First there are letters to the editor of the local newspaper declaring that the new commercial activity will destroy the amenities the community thrives upon (and a somewhat nasty response from the president of the local Chamber of Commerce to the effect that the community claimed, before this proposal, to want new industry—and now, the president laments, there are only complaints when a corporation offering jobs wants to move into the area). A meeting is called at the local civic center to discuss the perceived problem. Participants speak heatedly about the proposal, all stating how it will affect their lives and the values they hold, attacking the landholding company as greedy, irresponsible, and so on. Two out-of-work

lumbermen respond angrily in support of the cutting but, greatly outnum-
bered, they stalk out of the meeting. The remaining participants agree that
the clear-cutting proposal constitutes a severe threat to their community.
A grievance has been perceived. This stage formalizes the essentially reactive nature of most environmental action.[24]

After a period of heated objections to the cutting, the president of a local
environmental group stands up and says: "Enough talk, we need to act
quickly. Let's form committees." After more discussion, three committees
are formed. One committee will contact state legislators to explore the possibility of using government funds to purchase the land or at least a buffer
zone along the river. A second committee will contact an attorney to determine whether an injunction might be obtained, based on existing legislation protecting streams, to prohibit cutting within 1000 feet of the river.
A third group agrees to mount a public education campaign to rally support
behind the cause. Once the committees are formed and begin their task,
we can say that, in this second stage, an *action plan* has been devised.

In the process of forming an action plan and in the first steps of committee work, a third stage will occur; the committees will *identify constituencies* to which they can appeal—legislators, judges, community groups, and
businesses that might be adversely affected. Most likely, the local groups
will form or make use of existing ties to larger environmental groups such
as the Sierra Club or the National Audubon Society.

As the constituencies are contacted and pitches are made, *justifications*
for particular actions will be employed. Some of these will appeal to economic and prudential concerns, others will be strongly moral in tone, and
some will use legal arguments. Of the moral justifications, there will be
appeals to individual rights of downstream owners (which might be legal,
moral, or both). The wildlife advocates may prepare a brief pamphlet on
the expected damages to wildlife populations, closing with an appeal to
rights of animals to an unspoiled habitat. And thus, the fourth stage of the
action process involves developing arguments that are likely to be effective
in causing important constituencies to support remedial action.

Finally, in step five, one or several *actions* will be taken. There may be
an injunction sought, a community-wide rally may be held, and so forth.
These actions may or may not succeed in redressing the grievance. Since
the first four stages may describe a sequence followed by multiple individuals and groups, several constituencies may be courted, using many justifications, and several action plans may be undertaken by different groups
in response to the same grievance. Coalitions may develop and cosponsored actions may evolve. In the end, compromises may be struck between
groups favoring total preservation of the area and other groups who place
heaviest emphasis on protecting the river itself.

Several observations are in order about the role of values in this process. First, values operate all through the process—from the original perception of the forest coming down as a problem worthy of community concern until the final signing-off on a compromise agreement or a clearcut victory for one of the parties. Values permeate the entire process.

Second, the values concerned citizens, including leaders, act upon probably do not represent carefully worked-out systems that are adopted after careful, rational thought prior to action, nor will actors always be able to trace their decisions to some well-formulated axiom. Actions and value discussions permeate each other, and concrete situations and choices may do as much to affect values as vice versa.

Third, there may be important differences, at the fourth step, between the personal values of advocates and the justifications chosen to influence an identified constituency. As Leopold and Carson did, our actors will tailor their public arguments to their chosen constituency.

Fourth, while individual values may play an important role at crucial points in the process, such as in the second step (discussions of a reasonable action plan) or the third step (as individuals volunteer to work on one or the other of the committees), individuals will hardly ever be *forced* to choose in any ultimate sense *between* value systems. For example, a family who owns a canoe rental downstream from the clearcuts might be avid wildlifers who insist that wild animals have a right to habitat—in this context at least this value will reinforce, not clash with, their professional and economic interest in protecting their business.

The action-oriented model presented here is not intended to minimize the importance of individual values in the policy process. It is intended rather to show how many roles values play besides determining choices between two clearly formulated and apparently incompatible courses of action. The action-oriented model also shows the extremely open texture of belief and value systems bearing on environmental actions. It emphasizes that the circular process described in the last section is not concluded prior to action, but occurs throughout the activist process. Consider again the family who owns the canoe rental. They are running a business and clearly see that a dirty river, laden with silt and debris from upstream cutting, is bad for their business. At the same time, their deep love for wildlife and the outdoors was what convinced them to leave a comfortable life in the city and build their struggling business. They will use economic arguments ("our local economy is based on recreation—the clear-cutting is bad for the community") and biocentric arguments ("the wildlife was here before us or the landholding company—the animals have a right to an unspoiled habitat") in tandem. They are unlikely to see these as conflicting commitments to two opposed systems of thought such as the Dominant Social Paradigm or the New Environmental Paradigm. They will, rather, see their different arguments as complementary routes to the same conclusion. That philosophers and social scientists see anthropocentric utilitarianism and biocentric egalitarianism as directly conflicting systems of thought will no doubt seem to them a rather irrelevant abstraction. They would simply respond: "Look, the wildlife along this river is adapted to clean water, water that is maintained in this condition by the forest cover upstream. Our business, recreational canoeing, is based on the same clean water. There is no conflict between our use and wildlife use of the river. The conflict is with the greedy upstream landowner who wants to destroy the conditions that make the river good wildlife habitat *and* a good stream for canoeing."

Consider briefly what would be required to convince the canoe concessionaires that they are contradicting themselves by using both arguments in tandem. Suppose that the landholding company distributes a benefit-cost analysis in which it is shown that, from the human, economic viewpoint, the total value of the timber to society, plus jobs for lumbermen, and so on, outweighs losses to owners of recreational facilities, additional costs to local residents who will have to drive further to find attractive recreational sites, and so on. If such a complete benefit-cost analysis were offered and if that analysis were accepted as complete and accurate by the concessionaires, they might be forced to admit that the anthropocentric, utilitarian framework favors clear-cutting, that the biocentric perspective and the anthropocentric one "conflict," and that they must choose between the two viewpoints.

But notice what huge *ifs* this recognition of conflict involves. In fact, of course, the concessionaires will attack the benefit-cost analysis at numerous points. They will, for example, insist that most citizens place a high value on just knowing that pristine areas exist, intact with wildlife, and that this value was ignored in the analysis. They might also object that the analysis places equal weight on financial benefits to far-away city dwellers (who will pay slightly less for lumber) as on the economic well-being of the local community, which depends heavily on income from recreation. They might well object that, while a purely *economic* analysis favors timber-cutting, this ignores fairness in distribution. "It's *our* river, why should it be sacrificed for cheap housing for city-dwellers?"

My point is that conflicts between the abstract, monolithic systems of "anthropocentric utilitarianism" and "biocentric egalitarianism" are very difficult to pin down in everyday discourse about environmental quality. Such conflicts, if they exist, can be revealed only if one makes a very large number of contrary-to-fact assumptions: The company has offered a *complete* benefit-cost analysis, the company's opponents *accept* it as complete and dispute none of the figures assigned any of the values, the disputants all agree to forgo any objections based on distributive justice in favor of a pure calculation of society-wide economic well-being, and so forth. In short, the conflict becomes apparent only after the facts of the case, as in the fortress example, are artificialized and constrained. Given such constraining assumptions, disputants might be forced to admit that they are appealing to two conflicting values. Lacking agreement on such assumptions (as we always are in practice), however, disputants will have the option to insist that *both* value systems, if properly applied, point toward the policy they favor.

The open-textured model proposed here differs from that of the closed paradigmatic analyses in several respects. First, it is an action-oriented model, not a model designed to evaluate reasoning processes taken out of their context. Second, it does not assume that environmental activists *initially* develop a well-organized system of beliefs and values and *then* apply this system to determine what they should do in a particular situation. It assumes, rather, that most environmental activists perceive a grievance or opportunity and react to it—values and beliefs determine their actions, but these values and beliefs do not exist in a nicely organized and formalized system purged of all ambiguity and potential contradictions. The initial set of beliefs and values that an activist takes into an action-oriented situation will be used, developed, refined, and altered in response to a changing situation demanding action. Third, the factual component of the open-textured system of an environmental activist's beliefs and values is given more importance on the proposed action-oriented model. The canoe concessionaires use two quite different systems of value in tandem because they believe that, in the case at hand, a properly computed, anthropocentric economic analysis will dictate the same policy as the biocentric perspective. And, along with Leopold and Carson, they will feel justified in using either language and either worldview.

Part Two

Environmental Policy Objectives

6

The Pressures of Growth

Environmentalism and Elitism

Critics of environmentalism have often charged that the movement is "elitist." By this is meant, among other complaints, that environmentalists are mainly members of the middle and upper classes who have achieved a comfortable level of economic well-being and who want to "lock up" natural resources, discourage economic growth, and withhold upwardly mobile job opportunities from less privileged economic groups in the society.[1] Environmentalists, of course, dispute this criticism, arguing that it is unsupported by any reasonable interpretation of either environmentalists' goals or the socioeconomic data. Nevertheless, the criticism strikes a sensitive nerve. It is interesting that the charge is directed at environmentalists, a majority of whom are liberals or progressives,[2] both from the right, which claims environmental regulations choke off economic opportunities, and from the left, which argues that skirmishes over resource policy represent just one more episode in the ongoing war between the classes.[3]

What is undeniable is that the growth issue is the most difficult one facing environmentalists today. Here is a real dilemma. If environmentalists embrace economic growth in America, they apparently embrace endless sprawl, boom towns, high energy use, degradation of watersheds and wetlands, more chemicals—evils without end. If they *oppose* growth, however, they appear to favor unemployment, reduced wages, and economic stagnation.

About growth, the dilemma encourages ambivalence and waffling: In 1977, the Rockefeller Brothers Fund published *The Unfinished Agenda: The Citizen's Guide to Environmental Issues,* which emphasized a need for "a major transformation in human values" and argued that the United States has "enjoyed a development that is no longer possible for most [nations]." The United States must, the report urges, aid in "the transition from abundance to scarcity" and provide examples of how, "in a 'Conserver Society,' quality of life can be preserved (and, for many, increased) in an era of scarcity."[4]

In the years since the Reagan antiregulatory revolution, however, environmentalists have also emphasized the importance of economic growth in achieving environmental goals. In a 1985 agenda document (environmentalists *love* to compose agendas), the Group of Ten (chief executive officers of ten leading environmental organizations) said: "Continued economic growth is essential. Past environmental gains will be maintained and new ones made more easily in a healthy economy than in a stagnant one with continued high unemployment."[5]

This discussion has remained friendly (just barely), but hardly polite. For example, when I asked David Brower his opinion of the Group of Ten Agenda, he said, "They're chicken," implying that the current leaders of the mainline environmental groups are unwilling to take on the issue of growth, unwilling to do anything that might disturb the long economic upturn that characterized the 1980s. There is no question that environmentalists face a dilemma regarding growth. A more interesting question is whether the split is mainly political, involving strategies necessary in political maneuvering, or mainly ideological, representing a disagreement about the *goals* that environmentalists should pursue.

When environmentalists' agendas exhibit apparently opposing views about growth, it is worth wondering whether the term is used uniformly in all discussions. Since environmentalists employ a variety of worldviews, it is quite possible for disputants to "talk past" each other even when they are ostensibly using the same terminology. Being mainly reactive, environmentalists have defined their own worldviews in the larger political arena in which they do battle with commercial and development interests. It is important to understand that they define their positions on growth in opposition to politically opposed groups, as well as in opposition to each other. The most extreme opposition to environmentalism comes from unapologetic exploitationists, growth-oriented economists such as Herman Kahn and Julian Simon. These economists, like early American exploitationists, accept the Axiom of Abundance, doubting that there will be any shortages of raw materials caused by inefficiencies of production. Whereas this Axiom was explained in frontier days by reference to the vastness and apparent inexhaustibility of the nation's forests and farmlands, contemporary exploitationists rely on belief in the substitutability of one resource for another.

They believe that, if one raw material becomes scarce, prices of products manufactured from it will rise, thus stimulating innovation and discovery of a substitute product made from more plentiful materials. Exploitationists therefore see no limits to growth and favor policies that stimulate economic development because such development will provide the wealth necessary to fund research into new products substitutable for those manufactured from declining stocks. Exploitationists also believe that population growth presents no problem; Simon, for example, argues that rapidly growing populations stimulate economic growth and should therefore themselves be considered a resource.[6]

It is in response to developmentally oriented attitudes such as these, and to less extreme versions of capitalistic expansionism, that environmentalists argue for constraints on commercial development of natural resources. Three case studies—forest service policy, the "soft-path" approach to energy policy, and population policy—illustrate an emerging policy consensus.

Three Case Studies

FOREST SERVICE POLICY

Pinchot conflated his own career goal, to institute and lead an active, effective, and productive forestry profession, with the public goals of resource conservation. Following Pinchot's lead, the wise-use wing of the environmental movement initially excluded aesthetic and other "noneconomic" goals from its category of uses. Concern for fair and farsighted resource use, as it was institutionalized under Pinchot's influence in the governmental bureaucracies, centered on narrow commodity-oriented values. The bureaucracies therefore failed to develop a conception of "resources" broad enough to comprehend Muir's aesthetic/spiritual concerns.

Muir had equal difficulty explaining how wise use of forests for productivity fit into his more preservationist approach—he discussed these matters less and less as his opposition to Pinchot's narrow utilitarianism grew. Muir's followers and Pinchot's followers increasingly talked past each other, unable to comprehend the other's viewpoint within their own worldview. In the early environmental movement, no positive program of environmental policy emerged beyond shared opposition to unrestrained exploitation; no common moral and aesthetic vocabulary evolved, much less a shared vision of what it would be like for humans to live *in* true harmony with the world of nature, even while living *on* nature's resources. Leopold made bold steps toward a unification of these approaches but, even today, the environmentalists' dilemma, reflected as a lack of common vocabulary and common positive vision, affects discourse on resource policy.

In the intervening years, however, the lines of separation between the policies of the two traditions have blurred. For example, during the 1920s the Park Service—created as a counterforce against the orientation toward productivity in the Forest Service—succumbed to boosterism and emphasized development and road-building. Meanwhile the Forest Service, under the influence of Leopold and Arthur Cathart (who was hired as a consulting "recreational engineer"), set aside wilderness and roadless areas in the National Forests. Preservationists, therefore, allied themselves for a time with the Forest Service in opposition to the Park Service. By the end of World War II, however, Pinchot's emphasis on productivity had reasserted itself in the Forest Service, and the process of designating Forest Service lands as wilderness virtually ended. The Forest Service once again pursued a policy of unrestrained road-building and cheap timber sales through the 1940s and 1950s, until the revolution in consumer tastes, especially a rapidly growing demand for outdoor recreation, once again altered the Forest Service approach.[7]

John McGuire, whose career in the Forest Service began before World War II and who served as Chief of the Forest Service from 1972 until his retirement in 1979, says: "A major change in the goals of forestry during my time in the Forest Service has been a growing acceptance of valuing nonmarket resources. These still cannot be readily analyzed in the benefit-cost sense, but these values are much more likely to be recognized by people in general today."

Furthermore, he believes, these changes affect policy:

> Now, if the American people want more of something such as timber or forage from the National Forests, the Forests Service goes to budget-makers and Congressmen and says: "Here are some of the costs that will be incurred." We then insist on funds to support recreation, to mitigate effects on wildlife, and to protect other values that may be threatened. If we get the added funds, we can probably offset the costs to these other values.

McGuire sees these changes as intimately tied with the important changes in public attitudes that occurred during the 1960s and 1970s: "The growth in the public perception of nonmarket values has in a way both led to, and supported, the growth of the second environmental movement that occurred during the 60s and 70s. And these changes have had a profound effect on Forestry."[8]

As Samuel Hays argues, concern had shifted from production to consumption, and consumer interest in amenities pushed conservationists in the government resource agencies toward a broader conception of the public good. For example, in 1960, Congress approved the Multiple Use–Sustained Yield Act, which mandated the Forest Service to consider recreational uses as well as timber production in the National Forests; four years later Congress extended the idea of multiple use to Bureau of Land Management lands as well. In that same year it also passed the Wilderness Bill, committing the government to "managing" wilderness areas without concern for productivity.[9]

Today, resource managers are forced to make every decision with an eye on the effects it will have on broader social values. Foresters who think of their profession simply as timber production are badly out of step—forestry has instead become a balancing act, a balancing of timber production against recreation, of forest management against wilderness values, of productivity of single-stand forestry against the goals of ecological diversity.[10] Foresters have therefore abandoned the assumption that they can manage forests for a particular purpose such as timber production while paying no heed to other social values.

Especially important is acceptance of responsibility among forest managers to protect biological diversity. Throughout the Forest Service there are popping up small cadres of researchers and managers who specialize in protecting biological diversity. By accepting this responsibility, the Forest Service opens itself to criticisms of its methods from outside the forestry profession. While foresters can claim to be experts on timber production, they cannot claim overriding expertise in protecting biological diversity. Whereas Pinchot's Forest Service could ignore outside criticism, insisting that their expertise gave them the right to manage forests without outside interference, today's forester is liable to be shown wrong by an academic biologist or a restoration ecologist if a management plan causes serious disruption in the composition of species in the forest.

This new responsibility for protecting biological diversity also forces forest managers to transcend single-species management. The early forays of foresters into wildlife management mostly involved attempts, such as Leopold's, to maximize one or a few game species, each with its own management plan. Over the decades, foresters have recognized the futility of atomistic, single-species management and have gradually accepted whole-habitat management as essential in most cases. Management for protecting biological diversity completes this trend—a forest manager, in the final analysis, is responsible for managing every species in the forest. This implies that the manager must be a forest ecologist as well as a timber manager.

Nor do I mean to suggest that all of the movement toward a consensus has been from the conservationist side. Preservationists, who once thought they could preserve natural systems by isolating them, have been forced to admit that even the largest parks do not constitute whole ecosystems, so wildlife management must range over park boundaries. But this quickly implies that the mix of uses around the park will determine the qualities of the park. This in turn implies that the parks must be seen as one element of the larger pattern of land use and that they can be protected only if that larger pattern is devised with an eye toward their protection from spillover of the activities surrounding the park.[11] Preservationists have therefore been drawn into debates about the proper complex of land uses around park boundaries and into society-wide debates about clean air and clean water. The preservationists, like the timber managers, have been driven toward management of whole ecosystems rather than isolated portions of those systems.

It is important not to paper over real differences. Attacks on the Forest Service by modern preservationists for below-value timber sales, for superfluous road-building, and for short-selling recreational values have never been more intense. Further, wildlife- and wilderness-oriented organizations fault the service for doing a miserable job of protecting diversity on the lands they manage. But the differences, however rancorous, are no longer all-or-nothing disputes. The Forest Service has clearly and explicitly accepted the legislative mandate that makes its management a balancing act. Likewise, preservation organizations such as the Wilderness Society and the Sierra Club have accepted the same formulation of the problem, accepting that the national forests will have some productive use, but arguing for wilderness wherever possible as a counterforce against a perceived bias of foresters toward timber production over other, aesthetic and moral goals.

The arguments now represent disagreements of degree and, as in the early days of conservation, these policy disagreements play themselves out as disputes regarding the appropriateness of particular uses for particular lands: What is the proper mix of economic management and contribution to noncommercial values for a public agency? While these issues are not easy, they are now discussed in a context of both sides accepting that management will involve finding a balance, and that the Forest Service will be held accountable for the effects of its activities on the whole range of social values. Essential to this consensus is the agreement that contextual management must supercede atomistic management—that management must take into account the important role of health and stability of the larger landscape in which productive and recreational units are embedded.

While this shift in values can be considered a victory of Muir over Pinchot, the victory is by no means complete. First of all, once a narrow worldview is entrenched in a bureaucracy, new legislation and broader leadership from the top will at best alter it slowly. On a deeper level, however, the methodological innovations of hypothetical market techniques for measuring amenity values represent a response to only one aspect of Muir's and Leopold's disagreement with the early resource managers. The Axiom of Usefulness performed two key functions in the early conservationists' worldview: It focused their efforts on commodity production, but reliance on the unquestioned value of maximizing sustainable production also encouraged resource managers to believe that their enterprise of "scientific resource management" was a value-free discipline. Since scientific resource managers merely maximized measurable outputs of desirable products, their professional task seemed to them to involve no value judgments.

The acceptance of aesthetic and amenity values has broadened the mission of resource managers, but many of them never questioned the underlying assumption that social preferences can be measured scientifically and can be used to determine what values public agencies should pursue. The broadening of values that took place in the post-Pinchot resource agencies therefore involved no attack on the assumption that economics, perhaps supplemented by questionnaires to determine nonmarket preferences, could scientifically determine the goals of the profession.

One might say that the Axiom of Usefulness was replaced with the Axiom of Consumer Value in their minds and in their management approach. The Axiom of Consumer Value, unlike its predecessor, countenances a broad conception of human value—a product or experience is valuable to the public if some members of the public desire it. But the Axiom of Consumer Value shares with its predecessor the implication that the task of resource managers is to maximize public satisfactions, and it views those satisfactions as "givens." Wide acceptance of the Axiom of Consumer Value among resource managers, therefore, represents a broadening of management goals to include aesthetic, recreational, and amenity values, but does not represent an acceptance of science as ecstatic and normative. Nor does it represent an acceptance of Leopold's conception of conservation biology and environmental management as a value-laden search for a culturally adequate conception of man's ethical relation to land.

ENERGY POLICY

Current recommendations of environmentalists regarding energy policy provide another illustration of how broad policy agreement has emerged against a backdrop of varied worldviews. Environmentalists today employ a distinction, coined by Amory Lovins, between two "energy paths," referred to as the hard path and the soft path.[12] The hard path, which is favored by large energy-producing corporations and by the federal government, extends current policy by encouraging centralization, capital intensivity, and large-scale and complex production systems, and generally focuses on increasing energy supplies to meet growing demands. The soft path, by contrast, is favored by environmentalists and urges decentralization and smaller capital investments in small-scale organizational structures. The soft path idea refers not so much to specific policies as to an approach: The energy problem is no longer characterized as a search for an energy supply adequate to meet a certain demand level determined by an assumed correlation with projected rates of economic growth. On the soft path approach, social needs and goals are analyzed, and energy sources adequate to satisfy these are proposed. By emphasizing energy production that is matched to specific, local needs, the soft path approach encourages conservation and a variety of power sources, including increasing reliance on renewable sources of energy such as solar, wind, and extraction of power from biomass. In general, these methods are directed both at reducing demand and at creating less complex generating capacities that are labor-intensive rather than capital-intensive.[13]

The hard path follows naturally within an exploitationist perspective—it assumes no real shortages and that dwindling fossil fuel supplies will be replaced by ever-greater and essentially unlimited nuclear power capacity. Indeed, the exploitationists' Axiom of Abundance was until recently expressed in confident predictions of future nuclear energy "too cheap to meter." Further, hard path advocates correlate energy use with growth in the gross domestic product (GDP). In the wake of the energy shortages of the early and mid–1970s, there were three major studies of energy futures. Each of these reports established future energy demand by projecting growth in internationally aggregated GDP's and matching this demand with a mix of energy supplies. While the studies assumed that growth in energy demand would be slower in the future than it had been in the past, the methods used to project demand assumed that growth in GDP required a given level of energy use. Here we see a corollary of the Axiom of Usefulness at work: Since production for human use is valued, energy demand assumed to be associated with such production is unquestioned. Consequently, having projected huge future demands, demands that could not be supplied indefinitely by petroleum, the reports concluded by proposing a huge increase in nuclear and coal-fired generating plants for generating large supplies of electricity.[14]

While this approach is favored by powerful interest groups and by the U.S. Department of Energy, there are many reasons to question its wisdom. The switch to coal as a major source of power for generating electricity has greatly exacerbated problems of acid precipitation, while nuclear power generation has lagged behind projections. Especially given the apprehension associated with nuclear power generation, fears of sabotage and proliferation of nuclear weapons capacities, and the intractability of the problem of safe disposal of nuclear wastes, the hard path approach seems less and less attractive. As Amulya Reddy, writing for the World Resources Institute, argues,

> The inescapable conclusion is that even if conventional global energy strategies could yield a system of energy production that is sustainable, the world of which the energy system is a part would become unsustainable. Thus, it is not enough to provide a solution only to the energy problem—it must either facilitate solutions to other global problems or at the very least not aggravate them.[15]

Energy policy, in other words, must be evaluated in its large social context.

Emphasizing the appropriateness of means to ends leads soft path advocates to stress that choices regarding energy production cannot be separated from choices about the type of society we want. Lovins sees energy policy as characteristic of policy questions more generally:

> Many [people] who work on energy policy and in other fields have come to believe that, in this time of change, energy—pervasive, symbolic, strategically central to our way of life—offers perhaps the best integrating principle for the wider shifts of policy and perception that we are groping toward. If we get our energy policy right, many other kinds of policy will tend to fall into place too.[16]

Indeed, he sees a new approach to energy problems as the most fruitful line of attack on the "serious structural problems"—"centrism, vulnerability, technocracy, repression, alienation, and the stresses and conflicts that they bring"—of industrial societies.

Nuclear generation and large-scale synthetic fuel systems are not favored on the soft path, because the huge subsidies necessary to create and sustain these systems drain resources that should be used to create sustainable energy sources. Thus, while it is admitted that it is technically possible to explore a broad range of possible options, including nuclear and synthetic fuels as well as solar and wind power, soft path advocates insist that we must, in practice, choose. If we invest heavily in nuclear and other centralized methods now, these will create vested interests that will compete with and stifle the development of safer methods based on renewable supplies later.[17] Soft path advocates therefore emphasize immediate conservation measures and immediate use of renewable energy supplies where feasible, and recommend supplementing these measures with temporary transitional methods, such as the development both of low-sulphur coal and of new technologies for burning high-sulphur coal without heavy emissions.[18]

Environmentalists have embraced Lovins's conceptualization of two alternative paths, and they have enthusiastically lent their support to the soft path approach. *An Environmental Agenda for the Future*, the 1985 agenda document quoted toward the outset of this chapter, criticizes the Department of Energy for emphasizing the need to increase energy supplies, saying their approach "is likely to cause environmental destruction, [and] economic disaster, and possibly even push the nation to the brink of war over Middle East oil." By contrast, they say, "The soft energy path outlined in this chapter would bring about a cleaner environment, fewer international tensions, and a stronger economy. The best energy strategy is to increase greatly the productivity of the energy currently used and to shift to a future based on renewable energy sources." [19]

Nor is the enthusiasm for the soft path approach of conservation and use of appropriate technologies limited to leaders of environmental groups. Robert Mitchell found in a 1978 mail survey that 82 percent of a random sample of members of environmental organizations approved of "alternative soft technologies" and 75 percent expressed "a great deal or quite a bit of personal interest in solar power." [20]

We see, then, a quite impressive consensus among environmentalists favoring a soft path approach to energy policy. Justifications for the approach, however, are more varied. The emphasis on conservation, on expending capital to improve efficiency of end-use technologies rather than on increasing energy production capabilities, and the gradual switch to renewable sources of energy where possible are all justified straightforwardly within a classic Pinchotist, wise-use system of values that emphasizes efficiency. Wise-use environmentalists argue that insulating houses, improving efficiency of gasoline engines, and employing other techniques to gain the same outputs from less energy can do more to close the energy gap than expensive investments to increase supplies. They carry this argument one step further by generally opposing the expansion of capabilities to use fossil fuels or nuclear power to generate electricity at central facilities, favoring direct use of fuels for space heating. Lovins argues that intensive electrification is grossly inefficient: "At least half the energy growth never reaches the customer because it is lost earlier in elaborate conversions in an increasingly inefficient fuel chain dominated by electricity generation (which wastes about two-thirds of the fuel) and coal conversion (which wastes about one-third)." [21]

Interestingly, the argument for soft path technology also partakes of Pinchotist populism. Lovins argues that, while large generating plants create jobs during construction, these are highly specialized and temporary jobs, while the soft path is more labor-intensive in the long run and therefore creates more permanent jobs. [22] He argues that the soft path is superior to the hard path in that the latter requires "trading off one constituency against another—unemployment versus inflation, economic growth versus environmental quality, inconvenience versus vulnerability—[but] a soft path offers advantages for every constituency." And, in a passage reminiscent of Pinchot's attack on the timber barons, he concludes: "Thus, though present policy is consistent with the perceived short-term interests of a few powerful institutions, a soft path is consistent with far more strands of convergent social change at the grass roots. It goes with, not against, our political grain."[23]

Much of the case for the soft path approach then, can, be supported by appeal to the wise-use, conservationist worldview—by an appeal that rejects the Axiom of Abundance but never questions the Axiom of Usefulness. The soft path is justified as providing the greatest material happiness *for the greatest number of people,* emphasizing the interests of the common people as against the wealthy interest groups, and the future as against the profligate use of energy for present profit.

But the argument for the soft path approach also questions the emphasis on consumerism that is implicit in the hard path approach: "The basic tenet of high-energy projections is that the more energy we use, the better off we are. But how much energy we use to accomplish our social goals could instead be considered a measure less of our success than of our failure . . . [and] how much function we perform says nothing about social welfare, which depends on whether the thing we did was worth doing."[24] Here, the criticism of certain material goals as not worth achieving is reminiscent of arguments that question not only the Axiom of Abundance, but also the Axiom of Usefulness and its assumption that all production for human use is a good thing.[25]

Others have justified the soft path based upon arguments that question the value of the modern human condition. Kristen Schrader-Frechette, for example, has stated that "modern societies, dependent as they are upon sophisticated energy sources and rising gross national products (GNPs), threaten to sever us from what is most human and most desirable in more primitive societies."[26] In modern societies, people have little time for the "sleeping, dancing, . . . telling stories, playing with children, and making music" that characterized the lives of preagricultural peoples.[27] In a similar vein, others, such as Murray Bookchin, Theodore Roszak, and Daniel Dickson, have argued that "the soft energy path ought to be followed because it is part of a decentralized government and society." Termed the anarchist argument by Schrader-Frechette, this justification of the soft path rests on the premise that "modern society, with its political and industrial centralization, is dehumanizing and oppressive."[28]

Environmentalists have, in energy policy as in forest policy, tended toward consensus. They have, in justifying this emerging policy program, appealed to a variety of value positions drawn from varying worldviews. On one level, this creates no serious confusion: There is no direct contradiction in following Pinchot in saying that the soft path provides a more efficient means to fulfill consumer preferences expressed by our society, while simultaneously hoping that the heavy consumerism implicit in energy demand projections will change. Efficiency in achieving currently expressed demands can be advocated as one part of a larger approach that also attempts to reduce demands.

On a rhetorical level, however, it is useful to note that these two arguments rest on divergent premises—to the extent that efficiency arguments propose less wasteful means to satisfy the expressed preferences of North Americans, they never question the Axiom of Usefulness. When soft path advocates insist that reductions in demand would represent a moral improvement in our society, however, they are embracing Muir's anti-materialism. A broad consensus on energy policy is explained and justified by appealing to basic values associated with divergent world views.

POPULATION POLICY

Attitudes toward economic growth are closely entwined with attitudes toward population growth. The exploitationist treats population growth as a resource to be encouraged, rather than as a problem. This policy approach depends directly on the Axiom of Abundance—if there can be no shortage of raw materials, there are no limits on the number of humans that can survive on the planet. Environmentalists who, like Pinchot, reject the Axiom of Abundance while never questioning the Axiom of Usefulness, see rapidly growing populations as a problem insofar as they outstrip the ability of the society to produce goods and services in an orderly and sustainable manner. On this worldview, emphasis is not placed on absolute limits, but it is argued that population expansion should be restrained so that growth does not exceed the potential of rational development to provide goods and services.

Like Pinchot, his followers recommend the rational and sustainable development of resources to provide for expanding human needs. They cite well-known statistics showing a close negative correlation between economic development and birth rates. Throughout the Western, developed world, population growth rates have declined as soon as the standard of living rose, literacy increased, and rates of child mortality dropped. The conservationists' worldview, then, focuses attention on development as the solution to the population problem. They recognize, of course, that rapid growth rates exacerbate poverty and exert pressure for unsustainable use of resources. They therefore advocate birth control measures as a means to allow rational and sustainable development until the day when improving standards of living will relieve pressures toward rapid and unsustainable resource development.

Since Aldo Leopold applied the ecological model to human population growth, an application that led to a concept of "carrying capacity" of ecological systems to support human populations, environmentalists of the preservationist persuasion have insisted that there is a limit on the number of humans that can be supported both for specific geographic locales and for the earth as a whole. Leopold was quite explicit in saying that these limits are relative, and he listed parameters governing the limits of exploitation of natural systems.

The uncertainties involved are clearly expressed in George Woodwell's essay "On the Limits of Nature":

> The earth is obviously finite; its resources limited. But the questions of how large it is and what the limits might be for the support of man are not as easily answered as the concept of a finite earth might imply. Complexity enters at every turn. . . . The question of the limits for the support of man, at first sharp and reasonable, loses clarity under scrutiny; what seems finite becomes infinite, and what should be infinite becomes finite. And the entire frame of references shifts with human adaptation to a new circumstance. Nonetheless, there are limits although the limits may not themselves be unitary, stable, and finite.[29]

Environmentalists, given an agreement that limits exist, however uncertain and flexible, have also gravitated toward a dual strategy for addressing the population problem. They believe that vigorous birth control programs should be supported. They believe, also, that international development organizations should target projects that will improve the standard of living of the poorest societies, especially the poorest members of those societies.[30]

In practice, then, both groups of environmentalists recommend pursuing both strategies simultaneously: to restrain population growth as much as possible in the short run by use of family planning methods, and to reduce population pressures by raising standards of living among the world's poor.

Appealing to different worldviews, however, creates differing rhetoric. Pinchotist conservationists emphasize development as a solution, while others emphasize reduction in birth rate through aggressive birth control programs.[31] There are, without question, differing positions about how much to emphasize each strategy, and nasty disagreements concerning the amount of coercion that should be used in the family planning approach have arisen. These disagreements no doubt derive from differing worldviews, but the disagreements on a policy level clearly involve matters of degree and emphasis and do not preclude environmentalists from working together to develop both a rational policy for limiting population and a plan for the sustainable development of the world's resources.

Environmentalism and Elitism Revisited

Environmentalists, then, have evolved toward a broad consensus regarding policy options in such important areas as forest policy, energy policy, and population growth. At the same time, because they have failed to develop a shared, positive worldview, environmentalists continue to employ differing concepts and arguments to explain and justify these policies. One of the most striking manifestations of this lack of a positive worldview appears in the ambivalent attitude many environmentalists exhibit toward economic approaches to resource analysis.

Economists and resource analysts can, in such cases as Forest Service Policy, counter the argument that they overemphasize consumption at the expense of other expressed preferences by expanding their analyses to measure more and more amenity values. But to the extent that they *accept* the current mix of consumptive and amenity values expressed by the society (and this is a methodological necessity if they are to base policy on those values at any given time), they cannot express the environmentalists' moral argument, for that argument essentially requires a *transformation* of those values.[32] If preferences are accepted as the basis for determining policy, they cannot be simultaneously questioned as overly consumptive. A system of economic analysis that merely records expressed preferences cannot, therefore, fully express the moralistic strain in environmentalism, the view that the preferences expressed by North Americans are overly materialistic and consumption-oriented. It is perhaps this inability that partially explains the deep mistrust the environmentalists hold for economists.[33]

Another source of this mistrust is the tendency of economists to use merely quantitative measures of growth, such as gross national product, in their discussions of "growth." Environmentalists, on the other hand, insist that it is possible and important to distinguish between "quality growth" and mere quantity of growth.

This case is made eloquently, and applied to third world problems, by Reddy:

> The experience of the last thirty years shows clearly that growth must not be equated with development. Far more important than the sheer magnitude of growth is its structure and content and the distribution of its benefits. Once a growth process benefiting the elite has taken place, the material output of that process (for example, processed and packaged foods, expensive cloth, luxury houses, universities, and private cars) cannot be transformed easily into growth for the masses (cheap food and cloth, low-cost housing, mass health care, education, and transportation). The scarce resources available for development can go much further in eliminating human suffering if targeted on meeting a limited set of well-defined basic human needs.[34]

David Brower, who has long had a reputation as an uncompromising preservationist, now emphasizes the distinction between growth through destruction and another type of "growth," which he calls "restoration." He says:

> This, to me, is the big question: the growth question. We have not ana-
> lyzed the costs of it. I keep asking two questions: 1. How much growth
> *must* we have? What we need is growth based on restoration; and, 2. How
> much growth can we no longer afford? We can no longer afford growth
> based on destruction. That, for me, is an axiom. The growing economy is
> our death notice. You cannot have an ever-growing economy on a fixed
> planet; and there's no way of getting off the planet that we can afford.
>
> I like growth; but not unquestioning commitment to "a growing econ-
> omy." Some people feel better if the GNP is growing by a certain per
> cent. In Japan, it's 8 percent. At that rate, by the time a child born today
> can vote, you need 30 times as many resources as you have now and by
> the time the person retires you need 300 times as many. Where the Hell
> are you going to get them? They say, we don't mean for it to go on for so
> long. But I say, "Stop. Stop while Japan is still beautiful."

When asked in what sense he favors growth, Brower replies:

> I believe in preserving—but if you're a preservationist, you're trying to
> hold the boundaries, and you get shot at. So I now emphasize restoration.
> Here's a place for good investment, if you have a good proposal. This is
> an investment in the future. And look at where the pay-offs come; and
> look at the enormous pay-offs you're going to have ten years from now,
> twenty-five years from now, compared to a rapidly diminishing pay-off if
> you grow through destruction. We need to present our position that way.
> When we do, it becomes exciting. . . . That's the new direction. If you
> want growth—I'll call it that because people love the word—we want growth
> based on restoration rather than destruction. There are jobs in it. There"s
> money to be made in it, and our children are going to like it, because
> they'll have something livable to build a house on when they can afford
> it.[35]

The complexity of the environmentalists' overall case—which questions the social value of indiscriminate growth, advocates growth tailored to needs, emphasizes the importance of equity issues as well as efficiency issues, and questions the value of consumerism itself—can obscure the solidity of their policy consensus. Environmentalists favor sustainable growth to meet essential human needs, they favor growth that creates lasting job opportunities, and they favor growth in services, as long as this growth contributes to the solution of social problems. At the same time they oppose the traditional, exploitationist conception of growth that pursues growth as an end in itself, uses raw materials in a manner that is unsustainable, and equates social welfare with the satisfaction of consumer preferences that are often created by media blitzes rather than true social needs. In this context, environmentalists sometimes object to growth—they are then, implicitly, accepting the concept of growth associated with the long-dominant exploitationist worldview.

Consider one instance of this objection to quantitative growth as a good in itself. Because unemployment is an important social problem, environmentalists can argue that growth in GNP, if it fails to create job opportunities, is of little social value: "Unemployment and inflation are only the first of a long list of distressing side effects of a high growth, high technology, high risk approach to our energy problems."[36] As the 1977 Agenda says: "Survival requires a redirection from things that are merely countable to things that really count."[37]

Environmentalists' underlying answer to the charge of elitism is now evident. Their critics assume there is an ironclad connection between growth in GNP and improved standard of living throughout the society—an implicit appeal to the trickle-down theory of economic development. Environmentalists dispute this connection and emphasize quality and equity in growth. It is therefore possible for them to attack unrestrained economic growth while insisting upon social justice and environmental stability.

Conclusion

It is now possible to summarize, in a small number of principles, the emerging consensus among environmentalists regarding the proper use of resources. In one sense, environmentalists oppose growth. They oppose the unrestrained growth that was a natural outcome of the exploitationist attitudes prevalent during the settlement of the continent. But environmentalists do not oppose all growth: They distinguish between growth in quantity of goods and growth in *quality*.

Quality growth can be measured according to three characteristics that distinguish it from mere quantitative growth. First, quality growth measures human benefits broadly, including amenities, aesthetic values, and nonconsumptive values that contribute to quality of life. Here, environmentalists have gone beyond Pinchot's narrow focus on material goods and, following Muir and early preservationists, have insisted that nonconsumptive benefits must be given equal weight with consumptive ones in environmental decisions.

Second, quality growth is judged according to its equitable distribution. Environmentalists value growth that helps the least-advantaged members of society, especially growth that addresses specific social ills such as poverty, unemployment, illiteracy, and pollution. Here, environmentalists speak from a long tradition—Pinchot, from the beginning, advocated the development of resources for all the people, not just for the wealthy. Nevertheless, the full integration of equity concerns into the environmentalists' agenda has been slow and halting. This failure, no doubt, lies behind the reticence of minorities and the less affluent to embrace the environmentalists' agenda. I have argued, however, that a commitment to quality growth can and should include a commitment to equitable distribution of the benefits of restorative growth. As these commitments are articulated more clearly in the form of support for improvement of quality of life for the less privileged in society, environmentalists will broaden their appeal and, it can be hoped, finally throw off the label "elitist."

Third, quality growth is sustainable across generations—it must not use up the "capital" stored in nonrenewable resources more quickly than substitutes can be developed, and it must not destroy the regenerative power built into the rich biotic systems that produce renewable resources.

On this third point, sustainability, environmental thinking has undergone deep changes since the days of Pinchot. Pinchot believed in sustainability, arguing that resources must be used to maximize benefits over the "longest period of time." But the Pinchotist criterion proved inadequate in practice, as was illustrated by the dust bowl, because productive biotic systems often produce regular yields for considerable periods, even while they are being stressed by overuse.[38] Lacking an understanding of the limits imposed by the functioning of the system involved, political pressures toward economic growth often press systems beyond their capability, even when yields appear steady from year to year.

In practice, Leopold's introduction of a functional, ecological model for judging resource use has had two main effects: First, it encouraged environmentalists to address population growth as an essential part of planning for resource use. One aspect of his land ethic was to foresee this consequence and to limit population by conscious adoption of moral restraint. This he saw as a matter of degree; smaller populations are preferable because they permit "less violent" alteration of biotic systems.

Second, the application of the ecological model, with its implicit acceptance that systems are stressed by unrestrained economic exploitation, amounts in practice to the assertion that there are *constraints* placed on the pursuit of economic goals in resource use. In essence, this implies that resource policy should be formed in two distinct steps. In the first step, ecological information on the biotic system in question, its strength and redundancy, its vulnerability to stress, and so forth must be considered. Some systems may be so vulnerable that they should not be exploited at all, while others may be extremely resilient and appropriate objects of heavy exploitation. Other systems range between these extremes, but the maximal degree of exploitation should be determined, according to the ecological model, by the functional relationships within the system and their resilience in the face of certain types of management.

The second step in deciding resource use can be taken only after natural scientific data, drawn from ecology, soil studies, climatology, and so on, have been used to establish the constraints on exploitation that are inherent in the land community. In the second step, economic considerations are used to determine which of the permissible modes of exploitation will maximize human well-being. Economic choices are therefore constrained by ecological data on the health of the system and its susceptibility to riskless exploitation. Economic criteria of resource management should therefore be used to choose among a constrained set of possibilities, possibilities that are constrained by a concern for the ecological health of the larger system in which resource-productive subsystems are embedded. In political practice, the two-step process is seldom so ordered. Political power supporting economic productivity usually forces environmentalists to agree to a compromise position—one that optimally fulfills essential ecological constraints, but also imposes minimal constraints on economic activity.

I believe that this two-step approach to determine sustainability—which first determines constraints by reference to natural scientific models of ecological communities and only then proceeds to apply economic reasoning—is the essence of the emerging consensus among environmentalists on resource use.[39] This is not, of course, to say that environmentalists have no remaining disagreements. Rather, it is to say that their disagreements can be understood more clearly if they are seen as disputes that arise within this consensually adopted approach to policy formation. Remaining disagreements can be considered of two types. Some disagreements are mainly rhetorical; these have no real effect on policy, but represent differences in conceptualizations and differences in values to which environmentalists appeal in supporting similar policies. When environmentalists differ regarding growth and development, they are often emphasizing elements from the variety of worldviews that have been used by environmentalists through the years. Similarly, some environmentalists emphasize population control more than development, while others emphasize the latter over the former—but both groups emphasize the need for stable populations within the limits of available resources, and both recognize that both family planning and increasing standards of living among the poorest members of society are essential steps to stabilize populations.

Other disagreements are more substantive and represent real differences in policy options proposed. Here, I would suggest that, while these differences ought not to be minimized, they appear mainly as differences in priorities and differences of degree once they are understood against the backdrop of the consensus I have outlined in this chapter.

For example, thrown on the defensive by the Reagan antiregulatory revolution, mainline environmentalists emphasized economic health as a political necessity, apparently abandoning the idea of a steady-state economy. A careful analysis of the idea of quality growth, however, suggests that advocacy of economic growth should be understood as growth that puts few strains on the larger environmental context. Concentration on this variable would recast the growth/no-growth debate among environmentalists as a matter of degree.

While one agenda group may implicitly accept the concept of growth appropriate to an exploitationist and decry "growth," and another may redefine growth and embrace it, both groups are arguing for "quality growth," whether they call it development within a steady state or growth through restoration. As was noted at the beginning of this chapter, the problem of growth, and of devising a national response to it, represents the greatest challenge of modern environmentalism. The point of this chapter has not been to claim that environmentalists have embraced a fully unified position on this difficult issue; the point has been more modest—that the distinction between commitment to simple quantitative growth and commitment to *quality* growth provides broad policy goals around which environmentalists can rally. I have no doubt that there will be heated debate about the exact nature and content of "quality growth" for years to come.

Growth through restoration, which evokes Leopold's idea that some forms of land use are harmonious with the contextual landscape, and protect and even enhance its "integrity," may well represent the banner behind which environmentalists rally. If so, they will emphasize a search, led by conservation biologists and resource managers alike, for a positive conception of ecological health. Restorative growth is growth that protects and enhances the health of larger, contextual systems.

7

Pollution Control

The Lady and the Chemical Industry

When active environmentalists were asked in a questionnaire, "Has there been an author who has most deeply affected your thinking about environmental issues?" respondents mentioned Rachel Carson about three times as often as any other writer.[1] Carson's book *Silent Spring* has been described as the primary catalyst in transforming the largely moribund conservation movement of the 1950s into the modern environmental movement of the 1960s and 1970s. Carson's work precipitated the shift from first-generation environmental problems of land and resource protection to second-generation problems, especially pollution, which were more pervasive, less immediately apparent, and in many ways more insidiously threatening to members of the general population. The rise of pollution problems to the forefront of public policy concerns required a new vernacular, a new way of speaking about environmental threats and solutions. Rachel Carson, it is said, succeeded in one place where Leopold had failed; she injected ecological concepts and ideas into broader public policy discussions. Her graphic writing style, as well as her considerable status as a successful author, succeeded in transforming public discussions of environmental problems into a more ecological context by emphasizing the ways in which persistent chemicals move through natural systems and into human bodies.

Immense economic stakes were involved in the pesticide issue; production of DDT, for example, quintupled between 1945 and 1962, as chemical manufacturers' sales climbed from just over $10 billion to almost $33 billion.[2] The publication of *Silent Spring* caused a huge public controversy; that controversy has set the parameters, as well as the tone, for much of the subsequent debate regarding environmental regulation and environmental policy.

Carson began her attack on the indiscriminate use of pesticides with "A Fable for Tomorrow," in which she described an imaginary town in the heart of America. She first described an idyllic scene of humans living in harmony with their surroundings, including woods and hedgerows inhabited by countless birds, and streams swimming with fish. But "then a strange blight crept over the area and everything began to change." Domestic animals died. Humans became ill. She invoked the metaphor of a silent spring: "It was a spring without voices. On the mornings that had once throbbed with the dawn chorus of robins, catbirds, doves, jays, wrens, and scores of other bird voices there was now no sound; only silence lay over the fields and woods and marsh."[3] The town described by Carson was imaginary in the sense that no place had suffered all of the disasters she described, but she insisted that the problems faced by the town were real.

In subsequent chapters, Carson cited scientific evidence of dangers to wildlife and humans from exposure to pesticides and buttressed her fable with accounts of real towns where careless use of pesticides, especially long-lasting ones such as DDT and the other chlorinated hydrocarbons, had killed wildlife and caused severe illness and death among humans.

Chemical manufacturers reacted immediately and strongly. After an abbreviated version of the book was serialized in *The New Yorker* magazine during the summer prior to book publication, Velsicol, a major pesticide manufacturer, threatened Houghton-Mifflin with legal action if they published the book. Chemical corporations and their supporters in the U.S. Department of Agriculture attacked the book, as well as launching huge public relations efforts touting the positive advantages of chemical pesticides. It is alleged that some chemical corporations threatened to withdraw advertising from magazines and newspapers that mentioned the book favorably.[4]

While some of the "arguments" used by the chemical industry were mere ad hominem aspersions—one attack questioned Carson's right, as a "spinster," to raise concerns for future generations—there were three arguable objections to the book: (1) the book was "emotional" and not "scientific," (2) the book and the policy changes it suggested would threaten the food supply, and (3) the book was representative of the thinking of fanatical nature cults who advocate protecting wildlife even at great expense to human welfare.[5]

With respect to the specifics of these charges, Carson's effort stood up remarkably well. She had checked her scientific statements with experts before publication, and disinterested scientists generally supported her data; even when they questioned some of her factual material, they were careful to indicate that those errors did not detract from the overall persuasiveness of her argument.[6]

Carson had covered herself against the charge that her ideas and policies would threaten the food supply. She said at the outset: "It is not my contention that chemical pesticides must never be used. I do contend that we have put poisonous and biologically potent chemicals indiscriminately into the hands of persons largely or wholly ignorant of their potentials for harm."[7] This point was reiterated several times, and one chapter was devoted to alternative methods of pest control. Carson's conclusions—that chemical pesticides should not be used where there are acceptable substitutes, that they should be applied to specific problems with the utmost care, and that less persistent pesticides should replace the persistent ones wherever possible—were actually quite moderate and certainly consistent with maintaining agricultural productivity.

Finally, while Carson used the fable of a silent spring to focus attention on the problem of pesticide use, a considerable portion of the book was devoted to threats to humans, not wildlife. She clearly believed that indiscriminate pesticide use threatened humans and nonhumans alike; to accuse her of nature cultism was just off-the-mark rhetoric.

Underlying each of these "criticisms" of Carson's work, however, was a less explicit but more potent disagreement. The real issue between Carson and the chemical manufacturers was one of basic values—Carson saw the problem of pesticide use as a *moral* issue, while the pesticide manufacturers insisted that it was a purely *economic* issue. In keeping with those early and formative debates, discussions of pollution policy have pitted Moralists against economic Aggregators.

The moral tone of Carson's argument was established in the first substantive chapter of the book:

> We have subjected enormous numbers of people to contact with these poisons, without their consent and often without their knowledge. If the Bill of Rights contains no guarantee that a citizen shall be secure against lethal poisons distributed either by private individuals or by public officials, it is surely only because our forefathers, despite their considerable wisdom and foresight, could conceive of no such problem.[8]

By referring to the Bill of Rights and by insisting on prior knowledge and consent, Carson was invoking a moral *right* to a minimally clean and safe environment. These rights were not open to negotiation, nor were they to be overridden by the economic and commercial benefits of pesticide use.

Chemical industry and agricultural interests, by contrast, insisted that the question of pesticide use is an essentially economic one concerning costs and benefits of various policies. They clearly believed that heavy pesticide use increased production and profits and that, if the question were conceived economically, it would be clear that any problems caused by pesticide use are more than compensated by their advantages in increasing productivity and economic activity. Thus, although Carson highlighted a specific problem, the magnitude of her effect must be understood in its larger social context. The tension between economic growth and environmental protection existed as a basic structural social problem: The publication of *Silent Spring* merely made explicit the underlying tension in American society between untrammeled economic growth and a growing concern for health and the quality of life.

Two Approaches to Pollution Policy

The divergence between the environmentalists' moralistic viewpoint and the economic viewpoint of industry and economists has resulted in two quite distinct conceptualizations of pollution problems and to two approaches to addressing those problems. Richard Ayers, Senior Staff Attorney at the Natural Resources Defense Council and Chairman of the National Clean Air Coalition, speaking in the tradition of Rachel Carson's moralism, describes pollution as a problem of "justice and fairness." "Pollution," Ayers says, "is somebody's garbage, somebody's unwanted material and polluters have no right to impose, involuntarily on me, an exposure to those materials for their profit."[9] We can think of Ayers, then, as a contemporary embodiment of the environmental tradition of Moralism.

This conceptualization of pollution problems stands in sharp contrast to one described and advocated by Milton Russell, Assistant Administrator for Policy, Planning, and Evaluation at the Environmental Protection Agency under Reagan: "To risk managers, and those comfortable with modern technology, pollution is an externality of production [a cost that is not priced in the process of production and distribution]; its reduction is a goal of public policy, to be pursued vigorously to the point that maximizes the *total* public interest."[10] Russell, who wishes to compute costs and benefits of all pollution control efforts, represents the tradition of Aggregationism.

Against the backdrop of these differing conceptualizations, there are also two approaches to remedies. One remedial approach, sometimes called the "command-and-control" approach, relies most heavily on legislatively determined regulations. This approach has led, in the desperate case of Los Angeles (where decades of regulation have failed to correct massive air pollution problems caused by an automobile culture expressing itself in a topographically defined basin that sometimes traps dirty air for days), to a

5500 page, 3-foot-high smog control plan.[11] The Los Angeles plan, which is unprecedented in its detailed control of individual and industrial activities, will require the reformulation of many products—paints, inks, glues, and deodorants. It will outlaw charcoal lighter fluid, regulate emissions from small businesses such as dry cleaners, and mandate switching from gasoline to alternative fuels in many automobiles. In short, the plan attempts to control pollution by regulating the activities that produce it.[12]

Important pollution control legislation written in the aftermath of the *Silent Spring* controversies has embodied the rights-based approach. The Clean Air Act of 1970 directs the Environmental Protection Agency to set primary standards for clean air to protect human health. Congress directed that uniform standards be set nationwide and that each air quality control region must reduce ambient air pollution concentrations at least to the levels decreed. Also, most important, it directed that the EPA set national standards at levels that "provide an adequate margin of safety to protect the public from any known or anticipated adverse effects associated with such air pollutants in the ambient air."[13] Similarly, the Federal Water Pollution Control Act Amendments of 1972 called for the elimination of all pollution discharges in navigable waters.[14] Ayers interprets this legislation as "articulating the view that no person, other than those who are inside buildings, should be exposed to pollution that represents a danger to their health" and therefore, as legislating a "right" to clean air and clean water.

In opposition to the rights-based approach stands the "balancing" approach. As Russell explains it, this approach is one of risk management, which proceeds in two stages. The first stage, that of risk assessment, involves a scientific finding that exposure to a particular pollutant presents a risk to health. In the second stage, risk management, the expert asks how great "is the investment necessary to reduce it, and does this investment make sense, given all the other demands on our economy?" Russell continues: "The rock bottom premise of the risk management approach is that not every good thing can be done. Resources are limited, and thus choices must be made among desirable ends." Russell advocates balancing the benefits derived from controlling a pollutant against costs and against all of the other benefits that might be purchased for the cost incurred. "In short," he concludes, "however regrettable, some environmental risks are simply not worth reducing, given competing ends, including other risk reduction opportunities."[15] This approach therefore eschews individual "rights," interpreted as the overriding entitlements of an individual to a healthy environment, regardless of economic cost.

The balancing approach has also been expressed in important legislation. For example, balancing the costs and benefits of various controls by the Environmental Protection Agency is mandated in the Toxic Substances

Control Act of 1976 and the Federal Insecticide, Fungicide and Rodenticide Act.

Because balancers see the problem as one of maximizing efficiency rather than as a problem of equity, they are drawn to an aggregative analysis of policy options; achieving the most aggregative risk-reduction per dollar cost is their goal. While balancers historically have supported command-and-control regulations in many situations, disagreeing mainly with Moralists regarding the stringency of standards pursued, recent developments have caused disagreements that cross-cut the old alignment of Moralists in the tradition of Carson and Aggregators in the tradition of Pinchot's economic resource management. Trained economists outside industry—perhaps because they have invented and developed the idea—strongly favor a move away from the traditional regulatory strategy toward incentive-based policies in pollution control. These new strategies are being advocated as the wave of the future because they inspire innovation and promise to reduce the costs of pollution reduction. Industry leaders, who usually have spoken the same language as economists, are understandably cautious, fearing that a switch to a new system of incentive-based programs—coming so soon after expensive equipment has been mandated—will require them to pay twice. Meanwhile, the Environmental Defense Fund has become an active advocate of the incentive program, and other environmental groups are showing increasing willingness to use incentive-based programs when they promise greater efficiency.

Accordingly, some environmentalists are speaking the language of Aggregators, arguing for economic efficiency, even as other environmentalists favor more stringent command-and-control regulation. Hence we find, ironically, some environmentalists—those Moralists who believe we must press for more efficient regulation—advocating the same broad policy framework as industry leaders, but for very different reasons expressed within a very different worldview. On substantive policy, of course, these environmentalists often disagree with industry about *how much* regulation there should be, and the shifting coalitions may be breaking down the dichotomy between Moralists and Aggregators in the discussion of pollution control policy.

Environmentalists have proven willing to experiment with incentive-based programs, and a fragile coalition of economists (who insist on the efficiency of incentive programs) and environmentalists (who want at least to experiment with incentives to see if they achieve more pollution control at less cost) has formed. Working with the executive branch after Bush replaced Reagan, this coalition supported the 1990 amendments to the Clean Air Act, which greatly expand the use of marketable pollution permits. These permits allow companies to reduce their pollution and then sell these reductions to other companies that find reductions in their own processes too expensive. Several other incentive-based programs are under discussion. One is "netting"—allowing companies to reduce pollution in one area to make up for new emissions in expanded operations. There is also experimentation with, or discussion of, several other approaches to incentive-based programs. For example, an imaginary "bubble" can be created over a plant or over a larger area, mandating reductions in the pollution load created within the bubble, and then letting the corporation find the best means to comply with the overall requirements. If the bubble encloses only one plant, this approach differs little from netting; if a larger area with numerous polluters is enclosed, however, it opens the possibility of a market in emissions reductions among inhabitants of the bubble.[16] The size of the bubble could then be designed to make regulation appropriate to local needs.

Notice that the dichotomy between Aggregators and Moralists does not appear in exactly the same guise here as it did in the area of wildlife protection and resource use. When the environmentalists' dilemma was introduced in Chapter 1, Moralists were portrayed as showing concern *for wildlife itself,* in addition to concern for human welfare, and this concern was understood as independent of human interests and capable of placing moral limitations on human actors in their pursuit of human welfare. Similarly, Moralists have approached problems of resource use by positing limits on human exploitation, limits that are inherent in ecological systems. The Moralists' position in resource use, therefore, seeks guidance from the study of ecology, and interprets that guidance as implying limits that exist independent of immediate human interests and are capable of restricting human behavior for noneconomic reasons. As the central controversy in pollution control policy has developed and unfolded since the 1960s, however, the Moralists position, while no less opposed to setting policy by applying an aggregative criterion based on human welfare, has appealed to individual human rights as the basis of limits on the pursuit of utilitarian goals.

Bill Butler, former Vice-President for Government Relations and Counsel for the National Audubon Society, who argued the case against DDT for the Environmental Defense Fund when they sought a cancellation before the Environmental Protection Agency in 1971–1972, recalls that he and his team

> "began by bringing forward all kinds of evidence on the devastating effects of DDT on wildlife, and it was by no means clear that we were going to prevail. When we got to the questions of carcinogenic, teratogenic, and possibly mutagenic effects of DDT, and experts testified that DDT was found in 99 percent of human beings at levels approaching those that cause

problems in experimental animals, that's when we got the attention of the agency.[17]

Butler says, "I think the major change in the 1970s was a broadening of focus from wildlife into the area of toxic chemicals. One can, of course, say that toxic chemicals have an impact on wildlife, but efforts of environmental groups are now directed at protecting humans from the effects of toxic chemicals." He recognizes that this change has carried environmental groups such as the Audubon Society far away from their original concerns for wading birds in South Florida. "If pressed, however, we can defend whatever we are doing in wildlife terms. Having said that, I think Audubon and other organizations are focusing increasingly on human beings and interests of human beings in interaction with their environment." Butler attributes this change to two causes. First, he believes that the EPA itself was constituted not as an "environmental" protection agency, but as a human health agency. As environmental groups tailored their arguments to affect the policies of EPA, they were therefore pulled toward taking a more human-oriented approach. A second factor, he believes, was pressure on environmentalists from civil rights activists, mainly blacks, "who saw wildlife concerns as protecting a particular suburban lifestyle." Environmentalists, sensitive to these criticisms, reoriented their arguments toward human issues that would include minority concerns such as clean air.

Butler's interpretation of environmentalists' "broadening" their concerns from wildlife issues to human health issues, however, does not neglect their concerns for wildlife. "The broadening of perspective amounts to viewing humans as wildlife—or as animals at the very least. In being concerned about animals, we're concerned about human beings, too."[18]

The Moralists' approach to pollution control is still clearly opposed to the cost-benefit balancing approach of the Aggregators, but it is now clear that Moralists are not always best seen as opposing the *anthropocentric* viewpoint of Aggregators. Their complaint might better be stated by saying that the Aggregators, by concentrating on human welfare as measured in preference satisfactions, are likely to ignore important constraints inherent in the ecological context. And, when the quality of that ecological context is degraded, there will be moral impacts on wildlife *and* on humans.

Toward Policy Reconciliation

As the Moralists and the Aggregators have conceived the problem of pollution in different terms and formulated different goals of pollution-control legislation, so they propose different approaches to remedies. One might suppose that no middle ground exists given the antagonism shown by proponents of the two types of arguments. In practice, however, the disagreements are somewhat less sharply drawn.

First, it is important to note that, while the opponents disagree whether pollution issues should be conceived in moral terms, there is no essential disagreement about what fundamental *values* a pollution policy should serve. As Terry Davies, former Assistant Administrator for Policy and Planning at EPA and a specialist on pollution legislation, says, "Advocates of economic balancing never question that their analyses should encompass all human concerns and values."[19] Aggregators do not deny that human health concerns are relevant, nor do they ignore wildlife issues. Their point, rather, is that all of these values *should be conceived as preferences of individual consumers.* They believe that, since all individuals in the society act from a whole bundle of personal desires and values, the best way to make these commensurate is to determine how much individuals are willing to pay to protect particular values, such as a risk-free environment, wildlife diversity, or consumable goods. Because Aggregators see pollution as a by-product of the productive society that makes everyone better off, they believe the proper remedy should reduce pollution only in balance with other values. They do not believe that the interest in reducing risk should be given special, overriding consideration in determining policy.

At the same time, economists do not deny that we live in a society ruled by laws that proceed from other approaches. Even balancers recognize that we balance against a backdrop of existing regulations. If a firm wantonly dumps chemicals that pollute groundwater and drinking supplies, it would be an extreme version of balancing that would impose no special penalties on the firm, but would only ask whether the community was willing to pay the cost of a clean-up. Accordingly, the balancing view must be understood as applying not to every case of pollution, but to those that occur within the limits of current laws. Likewise, it would be an extreme position that would deny us the right ever to impose new laws limiting the production or use of new chemicals that will predictably cause serious social problems requiring expensive clean-ups. The balancing view, therefore, is best seen as proposing a general approach to reducing risks in those cases in which *no criminal activity is involved* and in which pollution is a normal by-product of useful economic activities that, it assumes, should not be banned or restricted by law. Proponents of balancing, unable to deny that there are some cases of pollution that involve equity problems too serious to ignore, must advocate not the total elimination of regulation, but the replacement of regulation with economic incentives for cases in which efficiency concerns should override equity concerns.

Second, while it seems that the Aggregators are committed to a methodological principle that all values should be quantified in dollar terms so that cost-benefit balancing can be applied to public decisions, in practice they recognize that this is more of an ideal than a practical proposal. Paul Portney, Vice President and former Director of the Risk Management Program at Resources for the Future is a strong advocate of the balancing approach. Portney admits that the balancing approach cannot result in a simple calculus that determines policies; it still "forces the administrator to make very difficult decisions." Moreover, he says, "if all favorable and unfavorable effects were supposed to be expressed in dollars, so that precise benefit-cost ratios were required, it would impose a burden which economic analysis is not prepared to bear. In spite of recent progress in valuing environmental benefits and costs, the science is still far short of being able to make comparisons in a precise way." He concludes: "For this reason, the balancing approach is best left in a qualitative or judgmental form."[20]

On the Moralists' side, there is also a recognition that some form of balancing will take place. Ayers says,

"The question is not whether there should be cost-benefit balancing. It's done in Congress every day. Even if Congress articulates the standard as one where everybody is supposed to be safe, they are making this kind of cost-benefit balance. They're doing it at a number of rather subtle levels. Not only are they saying, "This is the appropriate goal," they are also making some surmises about how government really works. Edmund Muskie and his people were saying: "Look, we know perfectly well, when this law goes over to EPA, that there are going to be high-paid lawyers from the industrial side who are going to try to nibble it away. And they're all going to have fancy arguments and a million facts to persuade the agency. We want to make it hard for the Agency to accept those arguments, because we know the process is warped against the protection of the environment."

Thus Ayers believes that balancing does and should occur, but that it should be done by elected officials and should include "not just economic considerations, but political and strategic ones as well."[21]

The apparently sharp dichotomy between Moralists and Aggregators is also softened if one introduces a distinction between setting goals for a pollution policy and developing means to implement those goals. Moralists tend to emphasize the legislative setting of standards, which function to determine the goals of an administrative agency. When Congress sets a no-risk standard, as in the case of the Clean Water Act, these absolute standards are viewed as goals or ideals. "As clearly articulated legislative goals," Ayers says, "their function is to drive policymakers in government to work toward that goal, even in the face of tremendous pressures representing business."

Ayers explains the position of environmental groups with an analogy:

Imagine that the Federal Aviation Administration were to say, "We have decided that the appropriate goal for the Agency this year is for 450 people to die in crashes." Except for real ideologues, the reaction to this suggestion would be outrage. The skies should be safe. We know we are not going to eliminate crashes, or deaths from pollution, yet we believe that the appropriate policy to task the relevant regulatory agency with is to come as close as possible. And, if they're going to balance, we want them always to feel guilty about it, because we want to drive them toward ensuring greater safety.[22]

This distinction between setting standards and means to implement those standards makes available an attractive middle ground for environmentalists: Environmentalists can claim that, in the setting of standards, we should be Moralists. Every person has a right to a healthy environment. Just as our justice system strives for perfect justice, but falls short, our pollution policy should strive for perfect safety, even as we recognize that it will fall short in many cases. When the focus, then, is turned to questions of implementation, environmentalists might join economists in advocating market-incentive approaches as the best means, or at least as one legitimate means, by which we struggle toward perfect safety.

Accordingly, Ayers describes the current system as "not inflexible":

Congress has allowed flexibility in the achievement of the standards. They understand that standards protecting all human health are not going to be achieved immediately, in every instance. As an accommodation to this reality, extensions are permitted. When the standards are not achieved within timetables set, it's not implied that cities or industrial establishments will cease to exist after that, but that further extensions require that Congress make the decisions. Just because EPA can't give an extension doesn't mean it can't be given. . . . Congress has not said, "You must do the impossible." They've set a goal and they believe the standards should be set on that basis. The program will be adjusted, as necessary, with Congress doing the balancing, as they see the costs and benefits or the weight of interests involved. They've adjusted the program in a number of ways. They've given time extensions, exceptions for particular industries, and so on. All of these are messy from the point of view of economists. But they're not crazy. They are often rough approximations of a wise balance, achieved through political means.[23]

Portney, likewise, recognizes the usefulness of a distinction between ends and means, between standards and implementation. He has recently been quoted as saying, "Look, we're willing to let the societal, environmental goals be chosen politically or religiously or by reading the entrails of ducks. . . . However, we're going to try to see to it that economics gets used as we make decisions about how we're going to meet those goals."[24] In short, if environmentalists and Congress insist on using other than economic criteria to set standards, *at least* they should let economists search for the most efficient means to implement those standards.

Both sides agree that (1) all human values should be considered; (2) at least in some situations, legal restrictions will be required to protect rights of individuals against threats to health; and (3) flexibility in choosing means to implement standards is essential. The two sides disagree, however, regarding *where* in the political process important balances should be struck.

Political Accountability

Aggregators believe that Congress should pass very general enabling legislation and leave implementation up to experts at the Environmental Protection Agency. Portney, defending balancing at EPA, says:

> It's clear that, under any EPA administration, balancing is going to take place. It was done under Dave Hawkins during the Carter Administration (which is about as liberal as you're going to see the Air Office run); it was done by Joe Cannon in the Air Office under Reagan. That is one reason why I want to see balancing made explicit in the legislation. It's going to be done and everyone knows it. The liberal Democrats are going to place a higher value on clean air and a lower value on economic dislocation, and conservative Republicans will do the opposite. If we make it explicit, we can all see what values are being attached to these costs and benefits.[25]

Ayers, the Moralist, agrees that accountability is the issue; in contrast to Portney, however, he believes Congress, as a democratically elected body, should do the balancing:

> Balancing should occur in the process of standard-setting in Congress. If Congress concludes that that is the proper way to set the standard then the EPA should follow that standard. But the present situation is that Congress has been through this exercise and has decided that the appropriate goal is to protect citizens' rights to clean air. But the EPA has decided they're not going to carry it out. They're going to carry out their own goal, based on balancing. This, to me, is a very dangerous tendency. In the case of the Reagan Administration, the President was enlisted in the process.

In this form, emphasizing political accountability, the argument between Aggregators and Moralists has interesting parallels to the amateur/professional split personified by Muir and Pinchot. "The advocates of economic and technical means of setting standards do not understand that Congress may wish to avoid a government of experts," Ayers says, "because they fear that there will be less accountability and that industry will have inordinate effects on procedures like standard-setting if it takes place among experts within the agencies." At this point, the debate, although it remains on a reasonably friendly level, turns more rancorous:

These administrators and "experts" are not democrats (with a small "d"). The economists and technicians would like a government by philosopher-kings. And they want to be the philosopher-kings. It annoys them that Congress doesn't see the advantage of this. But Congress prefers to set goals, because they have a much broader conception of how government works; they know that the decisions will, no matter where they are made, be political. And they believe it's appropriate for political decisions to be made in a political setting, and made by elected officials, not appointed ones.

Ayers, therefore, sees the issue in political terms: "Nobody is going to permit economists using cost-benefit sliderules to decide, based on their own calculations, that a whole industry is going to shut down. So, these decisions will be political. As a practical matter, this whole argument comes down to 'which branch?' Will it be the branch that answers to the people, or will it be the branch of the experts?"[26]

Economist Portney, however, sees the EPA as the appropriate body to balance and to set standards: "Opponents of balancing insist that standard-setting is political. What is the EPA if not a political entity? The President appoints the head of the agency; he appoints each of the program officers. EPA is not a nonpolitical agency. While they are not elected officials, they are accountable to elected officials."[27];

Portney also gives practical arguments:

> Congress should not set standards. Congress already has too much to do. Can someone tell me with a straight face that, in addition to everything else that Congress has to do, they should determine how much acryl-nitrate can be emitted from a dry cleaning facility? Is that what we elect Congress to do? The reason we have set up regulatory agencies is that the level of detail of the problems with which these agencies are concerned, and the issues are so technical, that it doesn't make sense for Congress to be worrying about them.

"There are problems with delegating these tasks," he concedes, "but the only thing worse than delegating them is Congress doing it itself. Congress is affected mainly by interest groups, political action committees, and so forth. Who believes that their congressman is really representing their interests when they [congressmen] write standards?"

Portney believes that maximal accountability will be achieved through the executive branch: "If the public does not accept administrators' priorities they can say to the President, 'We want a new administrator at EPA, or we're going to vote you out of office.' And it works. When Anne Gorsuch Burford and the other top administrators at EPA were not doing their jobs, there was such a clamor that pressure was brought to bear and they resigned."

Environmentalists, however, use a remarkably parallel argument to arrive at an opposite conclusion. The Group of Ten argues that the problem with cost-benefit analysis

> is that it tends to reduce the environmental debate to one involving technical numbers and thus becomes a debate that is difficult for the concerned citizen to understand, much less participate in. . . . Lacking the technical assistance to rebut every assumption contained in a cost-benefit analysis, the public will be forced to leave regulatory decisions to the "experts." Without public accountability, the experts have little incentive to make tough regulatory decisions.[28]

So the argument reduces itself to a question of accountability. Do cost-benefit analyses, conducted within government agencies, or political processes, conducted within the committee structure of Congress, afford citizens more input into decision processes?

While it may be needlessly cynical to say so, there is no small amount of professional self-aggrandizement involved in these debates. If major decisions are made in the policy analysis offices of EPA, it will mean more, and more meaningful, work for economists. If major balances are drawn in Congress, the lobbyists and attorneys associated with environmental groups will have more important work to do. To put the matter in such terms, however, *is* needlessly cynical, because the issues involved are indeed important and the questions of where the decisions are made and what conceptual framework is used to develop them are ultimately questions of *access* and *accountability*, of who will speak with the most effective voice in decision-making.

Avoiding the Dilemma

Pollution control debates illustrate that, while environmentalists *experience* a dilemma in addressing environmental problems, this is more a dilemma in choosing worldviews than a clear-cut choice regarding policies. Once one draws a distinction between setting standards and determining means to those goals, environmentalists can choose among a variety of general strategies for combatting pollution, each of which can be related to either worldview.

It seems reasonable, however, to argue that an extreme position favoring balancing—one that argues that *all* standard-setting and *all* implementation should be determined by cost-benefit balancing—is not an "environmentalist" position at all. If pollution efforts might be overridden by a social desire for more air conditioners, for example, environmental goals are given no special status distinct from simple consumer tastes. We can therefore limit the range of environmentalist responses to pollution to those approaches that, at least in some situations, limit economic activity in favor of a moral commitment to environmental protection.

Environmentalists, therefore, have consensually adopted the stance of Moralists to this extent—some basic pollution standards protecting human health or other important values must be understood as resting on moral premises and are not open to economic trade-offs. This is not to say that all environmentalists will agree on the exact formulation of the moral premises, or on the areas where they consider economic trade-offs and compromises to be morally acceptable. One area where there is considerable disagreement, it seems, concerns the effects of pollution on wildlife and the ways in which these effects should guide policy.

Given this minimal consensus, however, it is possible to characterize the disagreement between Aggregators and environmentalists who are at least minimal Moralists. Recall, for example, that Milton Russell argued that his aggregative approach to pollution control involved a two-step process, one of scientific determination of risk and one of "risk management." Balancing will occur in the second step; on that level Russell sees no moral constraints at all, but only a utilitarian calculus on which goals for pollution control must compete with every other social good for which consumers are willing to pay. Environmentalists would also adopt a two-step process, but one embodying two very different steps. In the first step, effects of a pollutant would be examined and it would be determined whether continued emission of that pollutant violated any moral rights (or other moral obligations, if any) of individuals to a healthy environment. If so, then standards should be set and penalties instituted to implement those standards. While instant compliance might not be possible in all cases, the goal of implementation should receive high priority and should not be overridden by a simple cost calculation. If no such rights were being violated, then further improvements in cleaning up the environment would be considered an economic preference in competition with other economic goods. This schematic framework, while it leaves considerable room for debate about what individual rights exist to a pollution-free environment (for example, is clear visibility a right or merely a consumer good?), sharply separates the environmental position from that of the pure balancers, who deny that rights should ever override economic interests.

An important continuing debate will concern the proper *means* by which morally supported standards should be achieved, as environmentalists argue to protect rights to a clean environment and balancers emphasize market-incentive approaches. Environmentalists would be well advised, I think, to remain flexible in this debate. The emphasis on equity considerations, appealing to the rights of citizens to freedom from harm resulting from air and water pollution, is justifiable in standard-setting. The manifest difficulties in proving a direct causal link between a specific act of polluting and subsequent harm to the health of an individual, however, should encourage environmentalists to seek the most efficient means to achieve a healthy environment, which might involve adoption of market-incentive techniques over command-and-control techniques in many cases.

As Carson saw, dealing with human health effects of pollutants inevitably leads back to dealing with ecological systems. Davies argues:

> If we must continually deal with pollutants after they get into the environment, they become a part of ecological processes. So you can't even deal with human health problems, viewed as multimedia exposures to chemicals which are shifted around and transported by natural processes, without dealing in a more integrated way with ecological questions about how pollutants are shifted in ecological systems.[29]

While this may not answer the concerns of environmentalists who feel we commit moral wrongs every time we release a pollutant into the environment and that pollutant kills individual birds or animals, it does begin to focus attention back toward the larger contextual system of nature. Even if our moral concerns regarding pollution are essentially human-related, they are also necessarily systems-related. Concern for human health includes a concern for a healthy context for human activity. And so it was that *Silent Spring,* which was dedicated to Albert Schweitzer, who advocates reverence for all life, led at first to an intense concern for human health, but eventually back to a concern for all processes of life.

8

Biological Diversity

Of Snaildarters and Mineral Kings

As a philosopher who has written on the subject of endangered species policy, I am asked from time to time to join a panel discussion on "the value of biological diversity." Consider a representative example: At the National Forum on Biodiversity, a 1986 conference organized by the Smithsonian Institution and the National Academy of Sciences, I shared the platform with three resource economists and one ecologist.[1] Everyone on the platform agreed that biological diversity has great value; the discussion focused on the question, can that value be quantified in dollar terms? I quickly perceived that I was in the middle of a polarized situation. The economists were there to demonstrate the efficacy of their methods for representing the value of wild species as dollars; the ecologist scoffed at these attempts as irrelevant at best and, at worst, as a symptom of moral depravity.

Hovering in the background of discussions like this are celebrated examples such as that of the snaildarter and the Tellico Dam. In that case, the Supreme Court halted work on an almost-completed dam because it would have flooded the only known habitat of the snaildarter, a three-inch member of the perch family.[2] The politically tortured case of the tiny snaildarter illustrates the dilemma environmentalists face in defending biological resources. Environmentalists initially opposed the Tennessee Valley Authority's plans to dam one of the last free-flowing stretches of the Little Tennessee River because it would destroy white-water canoeing, flood natural ecological systems, and destroy anthropologically important Indian burial sites.

Environmentalists made little headway, initially, as the bureaucratic processes ground forward and construction of the dam was begun. Then, in early 1976, in a dramatic development, biologists discovered a hitherto unknown species, the snaildarter, living in the waters upstream from the dam. Since the snaildarter spawned in shallow, fast-moving waters, the dam threatened to wipe out a distinctive form of life.

Environmental economists, anxious to use their quantificational tools, saw the Tellico Dam as a case in which assigning a dollar value to a threatened species might tip the scales in an aggregation of costs and benefits of proposed projects. Assigning dollar values to environmental damages such as a species loss, they thought, could act as a criterion to separate bad from good projects and policies. Environmentalists, however, were more inclined to see the snaildarter as a litigious tool, using the Endangered Species Act of 1973, which prohibits federal activities that may further threaten a species listed as endangered, to halt projects in the courts. In this case the environmentalists' dilemma appeared as two very different strategies to protect biological resources; here, ideology apparently implied a clear difference in action.

Economic Aggregators, by counting the dollar values of snaildarters, clearly implied that, in cases where those dollar values were insufficient to tip the scales in favor of preservation, they would be willing to let a species go in return for economic gains. Species were therefore conceived as resources among other resources. But the Endangered Species Act of 1973 contained, in Section 8, a categorical prohibition of any government activity that would further endanger any endangered species. Moralists believed that saving all species constituted a constraint on economic activity; no species could be bargained away for economic gain. Species protection, therefore, had a privileged status somewhat analogous to human individual rights, which trump mere economic interests. The generality feature of the Endangered Species Act could be supported only on the Moralist premise that all species have moral standing that is not economically negotiable

It was, interestingly, about this time that the question of "rights" of nonhumans became prominent in legal and moral philosophy, highlighted by Christopher Stone's much-discussed, serialized monograph, *Should Trees Have Standing?* A great deal appeared to be at stake. Business interests had initially blocked environmentalists' access to the courts, because environmental groups were unable to satisfy the operative economic criterion— demonstration of significant economic harm to members—needed to gain standing to sue to enforce constraints such as protection of species with no known economic value. Then, in a timely development, Stone's manuscript was placed in the hands of Justice William O. Douglas for review, even while the court was deciding the Mineral King case, which turned on the question of standing. The issue before the Court was whether the members of the Sierra Club had standing in the courts to contest Walt Disney Enterprises's plan to develop a ski resort on federal lands in the Mineral King Mountain range.[3]

Environmentalists lost the case, but Douglas cited Stone in his dissent and proposed that, if members of environmental groups lacked standing, perhaps the mountain range itself might have rights protected by law. Environmental groups could then assume the role, by analogy to adults' role in protecting the rights of children, of guardians of those rights. While the case was lost, there resulted a rich and lively debate on the question of whether nonhuman entities, from species to rocks, have rights.[4]

The Tellico Dam case presented no problems of standing, however, and environmentalists pressed their case in the courts. Since the Endangered Species Act declared that no government activity could further threaten any endangered species, and since the only known population of the snaildarter would be wiped out by the dam, the environmentalists had a strong legal case. The Tennessee Valley Authority fought all the way to the Supreme Court, but could not sustain its interpretation, that the law should not apply retroactively to projects already begun. The generality of the act was therefore confirmed by the courts—any project, no matter how valuable in human terms, must be stopped if it threatens *any* listed species. Work on the Tellico Dam was halted.

But this temporary victory in the courts only propelled the snaildarter back into political rapids. Development-oriented congressmen now had second thoughts about the Endangered Species Act, fearing the act as written might halt their own projects, and environmentalists faced the specter of an outright repeal of the Endangered Species Act. Environmentalists retreated by accepting a compromise amendment, cosponsored by the strong environmentalist Culver of Iowa. The amendment mandated a high-level government committee, quickly dubbed the "God Committee," that could decide, in the case of competing interests, when overwhelming national or regional interests justify the sacrifice of a species.

The God Committee quickly proved it would be no pushover for economic interests; in its first and most celebrated case, the Committee ruled that the Tellico Dam had from the beginning been based on a faulty cost-benefit analysis and that the dam, though 90 percent complete, should be left unclosed. The Committee therefore decided the case in favor of environmentalists without reference to rights of snaildarters or to the Endangered Species Act.

Environmentalists finally lost the Tellico Dam battle, but it was lost in

Congress, politically, not in the courts. Just as in the Hetch Hetchy case sixty years earlier, political maneuvering dammed a free-flowing river. Unlike the titanic struggle over Hetch Hetchy, however, this battle was settled behind the scenes, by the raw exercise of individual political power. Howard Baker, in whose district the dam would reside, slipped a rider onto an appropriations bill that passed an almost-empty Senate chamber after midnight. In essence, the rider, which survived the conference committee with the House, exempted the Tellico Dam from all federal regulations. The snaildarter, gasping for breath, landed in the hitherto friendly lap of the President. Carter, who had consistently opposed the dam, nevertheless signed the bill, perhaps because he needed the appropriations, but reportedly also because he needed Baker's support to bring Southern congressmen in line behind the Panama Canal treaty.

The Tellico Dam has yet to yield its first kilowatt, and it seems to be exacerbating water quality problems in other parts of the Tennessee River. The snaildarter, clever perch that it is, had not put all of its eggs in one branch of the river, however, and in one last, ironic twist, populations of snaildarters were subsequently discovered in other branches.

Environmentalists were not so lucky, however. Having put all of their ideological eggs in the rights-of-other-species basket, they found themselves in a politically nonviable position. They were forced to give up the position that all species must be saved, regardless of economic costs. The fall-back position that there is a strong presumption in favor of species preservation, and that a species should be sacrificed only if the human costs are overwhelming, could not be supported on either a utilitarian position (which implies such sacrifices whenever human benefits of a project outweighs its human costs) or on a rights-of-species position (which implies that no species should ever be sacrificed). Here is the environmentalists' dilemma at its most debilitating. Environmentalists, struggling along with fragmentary and overdetermined worldviews, must choose between utilitarianism—which does not prove enough to support their considered position—and the theory that all species have rights—which proves too much. Their policy position—that there is a huge burden of proof placed on any policy that sacrifices a species—while politically viable and apparently sensible, is left exposed, lacking moral and intellectual foundation.

The Aggregators and the Moralists

The dispute, as noted in the last section, between Aggregators and Moralists regarding biological resources is joined over the question, can the value of a wild species be represented in dollar figures? But environmentalists are not saying, when they oppose quantification, merely that dollar values assigned to particular uses of species are inaccurate or arbitrary. Their claim is the more radical one that the economists' whole conception of value is inadequate to reflect the value of wild nature.

Generally speaking, environmentalists and economists do begin by agreeing that wild species have use as commodities and amenities—as commodities in the form of alligator shoes and as amenities when we drive miles to hunt game or to view rare bird species—and environmentalists will at most quibble about the details of value assignments in these major categories. The problem with these categories is that few species, probably fewer than 10 percent, have economically significant commodity uses, and perhaps fewer have amenity uses for most citizens. Economists, recognizing the uneven distribution of these obvious categories of value, also recognize the *option* value of a species, which is the amount we, collectively, would be willing to pay to hold open the possibility that a given species will be discovered in the future to have important use as a commodity or an amenity. Since extinction rules out such future possible uses, it seems rational to pay some amount to hold open future options. Aggregators, as well as Moralists, therefore recognize that much of the value of species will manifest itself only in the future.

The problem with the Aggregators' concept of option values is that such values are, by definition, based on extreme uncertainty. For example, to illustrate how large option values might be, economists have estimated that the actual value of a particular species, *Zea diploperennis*, which was saved from the brink of extinction, turns out to have an economic value of $6.82 billion per year.[5] This nondescript grass, a member of the maize family, is a perennial, and promises to allow development of a perennial version of corn, creating tremendous savings in tilling, erosion reduction, and so on. Such examples suggest that we should be willing to pay considerably for holding open options, but the example is not very helpful in assigning option value to a species *before* much is known about it or its potentials. Once one knows a few important facts about the grass and its possible uses, one can estimate its economic value—but option value is not *this* value, it is the value people are willing to pay *on the outside chance* that a species for which we now know no uses will turn out to be useful. Option value, in other words, introduces a blank check into the value equation.[6]

Given the magnitude of the problems afflicting aggregative methods of valuing species, it is not surprising that Moralists ridicule the economists' attempts. Economists are very far from having, even for one well-known species, a complete accounting of all its present and future values. Given that many endangered species, especially in the tropics, have neither been named nor studied, the Aggregator offers an approach to valuing species that is at best theoretical.

The Moralists' position, however, is hardly more palatable. They insist that the true values of a species cannot be adequately expressed in dollar figures. Faced with an ongoing decision process in the executive-branch agencies that generates quantitative figures representing costs and benefits of proposed policies, Moralists, as conscientious objectors, shut themselves out of the decision process.

Environmental policy regarding biological diversity therefore seems to be impaled on this distressing dilemma—should preservationists play the game of economic analysis of policy decisions with one hand tied behind their back (because the economic framework of analysis creates only pale representations of the true values of biological diversity), or should they stay on the sidelines crying "foul" as the game of dam-building and habitat-destruction goes on without them?

These methodological quibbles and strategic disagreements are not the ultimate issues that separate Aggregators and Moralists, however; they are rather symptoms of a deeper ideological and intellectual disagreement. Environmentalists, following Moralists in the tradition of Leopold's land ethic, believe that human cultures have moral obligations to protect the health of *land systems*. They have not agreed how to express these moral obligations, but the shared commitment to a contextual ethic, often expressed as the organicist metaphor of the land as a living system, represents a consensus among environmentalists: They agree that one must protect the larger systems that are the dynamic, functioning environments and the spatiotemporal context of human activities.

Environmentalists can be distinguished from Aggregators mainly by their commitment to protecting larger systems through intergenerational time. This commitment to protecting systems explains the centrality of biological diversity in their policy goals. The sacrifice of a species represents the "violence" Leopold cautioned against—if a species is threatened, it is a sign that the system may be threatened with rapid and eventually uncontrollable change. This shared commitment to protecting the systematic level of biological diversity may be expressed in different value terms by environmentalists, because they share no common worldview. Simple aggregative approaches to valuing biological resources, however, cannot express environmentalists' commitment to protecting the integrity of systems and the processes that sustain this integrity through time.

To put environmentalists' moral point slightly differently, they are saying that a purely *social scientific* approach to setting biodiversity policy—a measurement of social preferences as expressed in markets—will not, given current social preferences, protect the *systematic* features of the environment. Citizens, acting as economic atoms, will support preservation of ecological systems only if those citizens are ecologically informed.[7] There is no place in a fully quantified description of consumer preferences to factor technical knowledge about ecosystem fragility, for example, into the decision process.

We can summarize this central, basic objection of environmentalists to the Aggregators' approach by emphasizing that environmentalists object to (1) the *static* nature of the economists' value analysis—they take no account of the *changes* in preference that would occur if citizens were educated about the ecological and systematic aspects of values of wild species; and (2) the economists' *atomistic* assumption that species will be given a value individually—their methods will not, therefore, capture those values that a species exhibits only when it is understood as a part of a larger system. Ehrenfeld, for example, objects that justifying the preservation of individual species because of their economic usefulness leads to a "piecemeal" approach to conservation and that this approach will fail for the very large number of species that have no known use.[8]

Thomas Lovejoy explains this concern more theoretically:

> The loss of a single species out of the millions that exist seems of so little consequence. The problem is a classic one in philosophy; increments seem so neglibible, yet in aggregate they are highly significant. Accordingly, endangered species with widely recognized economic value are likely to receive some consideration when threatened. But most endangered species are not of immediate economic value. Consequently there will be very few instances when a choice . . . between an endangered species and a development project will be made in favor of the endangered species. It is very difficult to marshal arguments showing that the particular endangered species will confer more benefits on humanity than the particular development.
>
> Yet if this reasoning is taken to its logical conclusion, most endangerd species will become extinct and the planet will be significantly impoverished biologically, with severe consequences for the welfare of people.[9]

Here is an objection to economists' procedures that stands quite independently of problems in assigning dollar values: If the species-by-species approach ignores important values that emerge only when species are looked at as parts of larger systems, the methodological assumption to evaluate species individually may cause a significant undervaluing of species. Species do not exist independently; they have coevolved in ecosystems on which they depend. This means that each individual species depends on some set of other species for its continued existence. A species may depend on just one other species for food, or it may depend on a whole complex of interrelated species. This seems to imply that if we take actions now that cause the extinction of any species, then the loss in future benefits should include losses accruing if any other dependent species succumbs as well. To extinguish a species on which two other species depend is to extinguish three species.[10]

The process of valuing species as a part of ecological systems becomes even more problematic as one recognizes that some species are "keystones" in their ecosystems. For example, when the Florida alligator populations dipped dangerously low about fifteen years ago, wildlife biologists noticed that many other species' populations also declined. They discovered that, during the dry winters in the Florida Everglades, other species depended on alligator wallows as their source of water. Accepting the economists' pledge to include a dollar value representing the total value of a species, the quantified total value of the alligators should include a contributory value nearly equal to the value of all the wildlife in the Everglades!

To determine the full value of any species, we would have somehow to determine the values of all of the other species that depend on it. To factor in these values, Aggregators would have to place an economic value on the whole ecosystem of which the species forms a part, determine the role of each species in supporting the ecosystem, and then apportion values to each individual species on the basis of its contribution to the entire value package. Unfortunately for the Aggregators' case, attempts to measure the economic value of ecosystems have fared no better than analyses of individual species.[11] Furthermore, this retreat would in effect be an abdication of a role, coveted by Aggregators, in deciding difficult species-versus-project disputes such as the Tellico Dam case.

Since Leopold's classic essay, "Thinking Like a Mountain," environmentalists have adopted a contextualist/organicist metaphor and rejected the atomistic approach to valuing biological resources. The economists' quantifications do not appeal to environmentalists because they do not reflect the contextualist nature of environmental problems, the sense in which a species must be seen as a part of a larger system. Nor can Aggregators recognize the special role of wild species in altering the worldviews of consumers. Since Thoreau and Muir, environmentalists have emphasized the central role of perception and experience of nature and the ecstatic aspect of scientific observation of wild species. The economists' methodological decision to measure the value of species in the static terms of expressed market preferences disqualifies them from expressing environmentalists' commitment to the dynamic and organic nature of human perception and valuation of nature.[12]

Saving What?

Life creates life—a puzzle and a paradox. Left undisturbed, forms of life reproduce themselves and multiply, creating a stunning variety of species and natural communities. As Leopold said in 1939, "The trend of [undisturbed] evolution is to elaborate the biota."[13] But this is no explanation—mere passage of time represents no causal force. Understanding the source of life's creativity represents the greatest challenge to modern biological science, just as protecting it represents the greatest challenge of environmental policy.

The idea of self-generation has had an important role in Western culture. Monotheism, a central idea in the Judaeo-Christian culture since the Jews began to worship Jahweh, is itself paradoxical: How could God exist eternally, and yet be creator of all things? The paradox was solved by concluding that God was responsible for his own existence. Were he not, it was reasoned, he would be dependent, un-Godlike. This idea subsequently supplied the ultimate explanation in the Great Chain of Being—everything else depends on God and is shaped by his will, but God depends only on himself. God must be a self-moved mover.

Today, the idea of a self-moved mover occupies a central role in modern biology. The study of the creative forces responsible for nature—the exclusive domain of theologians and philosophers in a world of fixed species created in seven days—has, since Darwin, been largely subsumed in biology.

The central role of self-generation in modern biological theory was illustrated for me once when I was speaking with an ecologist friend, pestering him to say something, without qualification, on the murky subject of the relationship between diversity and stability in biological systems. Finally, in exasperation, he said, "The only thing we can say for sure in this area is that diversity creates more diversity." This is the theory of Robert H. Whittaker, the late botanist and theoretical ecologist who worked at Cornell University. Diversity, he maintained, creates and sustains diversity.[14] It is an elegantly simple and yet paradoxical idea, and it holds the key to understanding the panoply of diversity embodied in natural systems.

So biology mimics theology. Its ultimate explanation is that life, wherever it came from, creates and recreates itself and other forms of life through death and reproduction. The vast force of nature is in this respect Godlike; the earth's community of life is a self-moved mover, and we are but one expression of a creative power outside us. This was the same insight—the recognition that we are but one expression of nature's miraculous creative powers—that struck Muir when he knelt before the orchid buds in Canada. He was driven to the conclusion that biology and theology are inseparable. Environmentalists still follow Muir in believing that science, especially biology and ecology, have this ecstatic aspect—the ability to transform perception and point the way to a more elevated outlook. While Leopold was slower to embrace mysticism than Muir, environmentalists in general agree that knowledge and appreciation of life in all its varieties and interrelations foster a sense of wonder, delight, and respect for nature, which in turn transforms the perceivers' outlook on life.

Here the scientific traditions and theological fascinations of modern environmentalists are united. Nature's complexity, its ongoing systematic processes, represents the mechanism of self-generation. Darwin demonstrated that creation is a continuous process; Muir insisted that this process must coherently be viewed as the working out of a larger plan of creation. Life produces life from death. The search for knowledge of the interrelations among elements of nature's system, therefore, represents the hope of environmentalists to understand who we are and where we came from; it represents also the faith that a scientific naturalism—a search for meanings and values in the natural context of human culture—can lead modern societies out of the morass of modern alienation.

For environmentalists, knowledge of the interrelatedness of nature supports an ecological conscience by encouraging a shift from a Cartesian worldview, which separates facts from values and posits a value-free science, to a worldview recognizing that values and facts are both generated within life processes. It is therefore these *life processes*, the creative complexity embodied in ecological systems, that environmentalists wish to protect.

It is important to distinguish between diversity and complexity. Zoos are more diverse than natural systems, but the species in zoos have not coevolved interdependent symbiotic, competitive, and predatory relationships with each other as they have in natural systems, because the relationships among the species are artificially limited by fences and cages. Indeed, zoos are not really systems, unless one includes the zookeepers and their activities of feeding, breeding, and culling. Diversity, in short, is one aspect of complexity, but it is really multilevel complexity that should be the object of protection.

As Whittaker analyzed it, diversity includes three levels: within-habitat, cross-habitat, and total diversity.[15] Species adapt to narrower and narrower niches along a limiting gradient, such as a variation in food source. For example, Terry Erwin of the Smithsonian Institution's Man and Biosphere Program extrapolated from species collected in Panama that as many as 1250 species of beetle may live in one square meter (10.76 square feet) of tropical rainforest canopy.[16] It is the complex relationship of the many interlocking habitats and microhabitats in the multiple stories of a rainforest—its cross-habitat diversity—that creates species diversity, not vice versa. The creative force is whole only in systems that can maintain their complexity over many generations. And thus it is the "total diversity" of a geographical region that is the most important measure of biological variation. It is a product of both within-habitat diversity and cross-habitat diversity. The complex, interrelated levels, changing in different scales of time, were referred to by Leopold as the "integrity" of an ecosystem: "A thing is right when it tends to preserve the integrity, stability, and beauty of the biotic community. It is wrong when it tends otherwise."[17] Leopold's viewpoint has been adopted by modern environmentalists: Protect as much biodiversity, including as much complexity, as is feasible, even if this protection incurs major economic costs.

Leopold's contextualist theory of management is supported by the scientific idea that trends in individual human actions can, depending on land fragility and other conditions, reduce energy flows through the large-scale systems we call "the environment." When accumulations of these effects stress systems, the stress usually manifests itself, first, as a gradual reduction in the redundancy of the system's energy pathways and, later, as an increased rate of change in the complexity and total diversity of the larger system. In worst-case scenarios, whole bioregions can be simplified, and become less able to resist drought and other extreme conditions, resulting in ecologic breakdowns.

To counteract these trends, Leopold recommended thinking in larger contexts, both temporally and physically. All human activities take place in a larger physical and temporal context that changes according to its own dynamic. In a healthy system, activities of individuals contribute to the larger system by creating or processing energy. Normally the larger system changes in ecological time and provides a relatively constant backdrop against which species compete. In a forest, for example, canopy trees are much more long-lived than the smaller plants for which they provide the micro-habitats, and provide an environing context in which these smaller plants compete. The behavioral and genetic adaptation of species can be understood only in the larger, temporal context in which populations wax and wane ecologically and evolutionarily. But human individuals, acting on much more rapidly changing *cultural* ideas are not inclined to think in mountain time, and economically determined decisions will change the forces of selection so rapidly that other species will fail to adapt and will disappear. The system will be stressed as more and more of its energy pathways will be broken.

The distinctive character of contextual thinking focuses the manager's attention not so much on individual actions as on their collective effects on the larger system and their effects on trends across more distant time. If Farmer Jones converts his woodlot into a wheat field, this may be ecologically significant or not, depending on the social and natural context of the decision. If Jones plants wheat, and Smith down the road lets his wheat field go permanently fallow, Jones's decision will probably be environmentally insignificant in the long run. Patchiness associated with variable disturbances is an important component of total diversity, and human activities can mimic natural patchiness as long as individual actions balance each other out. If Jones's woodlot is on a steep slope or if far more farmers imitate Jones than Smith over several years, however, the larger ecological context of the area will eventually be affected. Contextual thinking recommends that, as major trends in agriculture and land use unfold, environmental managers monitor the effects of these activities on their context. Contextual management does not treat all decisions to plant wheat equally; some decisions have moral impact because they are part of a trend that presses against a threshold, which varies according to the fragility of the systems involved. Activities that exceed local thresholds contribute to accelerated change that eventually destabilizes the broader system.

Leopold thought contextual management was necessary to avoid gradual or even rapid deterioration of larger systems when dominant actors with huge technological capabilities make decisions in human time. If these decisions constitute a definite trend, limits must be observed; steps must be taken to encourage individuals to break the trend in behaviors that threaten their environmental context. Contextual management, therefore, understands individuals as participants in larger systems, and examines the effects of trends in social systems on the "natural" systems that form their environmental context. The complex organization of ecological systems sustains energy flows among the diverse elements of the system and keeps the system functioning in a dynamic, but stable and integrated fashion. Conservation biologists and environmental managers, therefore, should proceed as physicians do—they should not simply treat one aspect or "organ" of the biotic system while ignoring the larger context, the organismic "body," in which that organ is embedded. Activities such as farming and game management that often involve alteration of elements—for example, a field or a farm, or a single species, or set of species that are desirable to hunters— of a system must not be undertaken without concern for the larger system in which the elements are embedded. Management, in other words, must not be just atomistic—it must be contextual as well.

Contextual management embodies concern for the three variables—the natural resilience of the land community, the density of human populations in relation to that resilience, and the rapidity and scale of the changes implemented. In essence, good environmental management shows concern not just for the productivity of the elements of the system taken singly, but it must also pay attention to the larger landscape in which they are embedded.

Since we do not yet know what activities are consistent with protecting the complexity and energy flow in natural systems, and since these will vary from system to system, we do not have a single "ideal" to guide management. The ideal must be worked out in a complex process that involves management experiments—but since we do not yet know what will protect and enhance the integrity of local systems, the search for such an ideal becomes part of the process. Conservation biology becomes a part of a social experiment, a part of the search for a viable concept of the good life for human inhabitants of the landscape. The human culture becomes, on Leopold's contextual analysis, a part of the ongoing, changing dynamic system of experiments in living in a natural context.

Hierarchy Theory

Recently, there has emerged within ecology a new and highly promising theoretical approach, "hierarchy theory," that bears striking resemblance to Leopold's communitarian model and contextualist approach to management. This new approach shows promise to give more precision to the concepts that plagued Leopold's theory of environmental management. Hierarchy theory is based on general systems theory and focuses not so much on the diversity of systems as on their complexity and internal organization—what Leopold called "integrity."[18] According to hierarchy theorists, natural systems exhibit complexity because they embody processes that occur at different rates of speed; generally speaking, larger systems (such as a community) change more slowly than the microhabitats and individual organisms that compose them, just as the organism changes more slowly than the cells that compose it. Further, the community survives after individuals die and, while changes in the community affect (constrain) the activities of the individuals that compose it, the individuals themselves are unlikely to affect the larger system because the individual is likely to die before the slow-changing system in which it is embedded will be significantly altered by its activities.

This is not, of course, to say that elements have *no* impact on the systems that provide their context. The elements, often called "holons" in the context of the analysis, are "two-faced"; each holon "has dual tendency to preserve and assert its individuality as a quasi-autonomous whole; and to function as an integrated part of an (existing or evolving) larger whole."[19] As a part, the holon affects the whole, but scale is very important—the "choices" of one element will not significantly alter the whole—but if that part's activities represent a trend among its peers, then the larger, slower-changing system will reflect these changes on its larger and slower scale. One cell turning malignant will not affect an organism significantly unless it represents a trend toward malignancy. If such a trend is instituted, then the organism might eventually be destroyed by that trend in its parts.

This multiscalar approach to time and space in ecology is reminiscent of Leopold's metaphorical discussion of differing scales of time and our perception of them in "Thinking Like a Mountain." Hierarchy theory provides a more precise conceptual model for what Leopold called "the land," which was for Leopold a slower-changing system composed of many faster-changing parts. He explicitly commented that our failure to see deterioration in the land community is due to our failure to recognize that ecological and evolutionary changes take place in a scale of time slower than that perceived by humans. Agriculturalists and game managers focus on the rapid-change systems that produce annual crops. The mountain, as Leopold explained metaphorically, must look at the value of wolves in a longer perspective.[20]

When nature is viewed hierachically, as a system of parts embedded in larger and larger wholes, holons higher on the hierarchy represent the environment of lower holons and constrain their activities.[21] Allen and Starr argue that "the positive aspects of organization emanate from the freedom that comes with constraint. The constraint gives freedom from an unmanageable set of choices; regulation [from above] gives freedom within the law."[22] According to hierarchy theory, then, natural systems are seen as organized units embedded in a hierarchy, with larger, slower-changing systems (the environment, or "context") determining the range of choices available to the smaller systems. This abstract, hierarchical model can be applied on many levels—cell/organ, organ/organism, organism/microhabitat, microhabitat/ecosystem, ecosystem/bioregion, and so on—and provides a sliding scale of concepts for analyzing relationships among the parts and wholes of living systems.

In ecology, which emphasizes relationships among systematic elements, hierarchy theory provides a tool for analyzing the multilayered complexity of natural systems, and shows promise to model the dynamic relationships among their parts. The hierarchical model may also point the direction toward a managerially useful concept of dynamic stability and ecosystem health. This concept will relate functions of faster- and slower-changing systems. The appearance of accelerating changes in normally slow-changing systems may indicate deterioration—illness or destabilization—in the land community. Viewed from above, the part is constrained by its larger, environmental context. Viewed from below, "normalcy" or "stability" is achieved as cells follow laws intrinsic to their behavioral capabilities—actions of one short-lived cell will normally cancel out the actions of others, achieving stability (understood as statistically probable regularities) on the larger systematic level.[23] The birth of an individual, for example, will not affect population stability if birth and death rates remain relatively constant. If, on the other hand, birth rates or in-migration rates soar, without corresponding increases in death rates, the trajectory of change in the larger system will be significantly altered or accelerated.

Leopold recognized that modern human societies, which combine astounding technological capabilities and unprecedented individual mobility, can rapidly alter trends that affect the larger context. Hierarchy theory can model these occurrences as exceeding parameters set by statistically understood patterns in the actions of parts. Radical, or systematic, change occurs when unbounded behavior of individuals within part of the system begins to cause accelerated change in the larger system.

Allen and Starr briefly discuss these large-scale matters affecting whole landscapes, emphasizing that changes recognizable at the individual level are normally "damped out" in the higher-level systems. They note that the Dust Bowl represents an example in which activities of individual farmers had little effect on the large-scale environment of the plains states as long as they occurred on a limited scale. Eventually, however, as more and more plainsland was converted to monocultural agriculture, a threshold was reached and the plains system, which had been changing slowly for millennia, underwent rapid change as a result of a pervasive trend in the behaviors of individual farmers.[24]

This explanation is isomorphic with Leopold's: The Dust Bowl resulted from the intensive and pervasive application of monocultures and intense grazing in a bioregion ill suited to intense use because of its cyclical patterns of rainfall and aridity. Similarly, Leopold, as a game manager early in his career, saw deer/wolves/hunters as a cell that could be managed, and he destroyed wolves to maximize deer for hunters. Experience taught him that, in fact, the cell must be considered a holon, and it therefore exists also as a part of its larger environment. Removal of wolves caused an irruption of deer; deer then overgrazed their browse, exceeding the capacity of the larger system to equilibrate these changes. A parameter (viewed from above, a constraint) is exceeded; the larger system becomes unstable, because redundancy is lost; under stress, rapid change such as dust storms and even desertification become more likely.

Hierarchy theory helps to explain Leopold's problem of parts and wholes, which we encountered in Chapter 3. Hierarchy theorists see their systems as aids to understanding—they are "epistemological," not "metaphysical."[25] Whether natural systems are themselves hierarchical is not the point, they say; viewing them as hierarchical makes their complexity intelligible to us. And yet O'Neill and his coauthors also recognize that "the task of choosing an appropriate system for investigating a particular phenomenon is inseparable from consideration of underlying organization and complexity."[26] Applications of hierarchy theory to environmental management, therefore, focus on the organizational complexity of the particular systems to be managed, and approach management problems by understanding them in the context of larger and larger systems. Environmental problems will be manageable only when they are approached on the proper scale. Hierarchy theory's pragmatic, epistemological approach can support the choices of ecosystem managers who are sensitive to context; a "whole" system for management should be one that fulfills both scientific and pragmatic criteria. Management goals along with a commitment to understanding the context of resource use and environmental problems ecologically, therefore, provide the boundaries of management—boundaries that are scientifically appropriate, given management goals.

Hierarchy theory may also point the way to a solution to the other problem that disabled Leopold's approach to management—the lack of an ecologically sound concept of dynamic health for ecological systems. A system is more healthy when impacts of human activities on larger systems do not accelerate change in the larger, normally slower-changing systems that environ those activities. Hierarchy theory, or some other model of complex systems, may someday furnish mathematical ratios of change across systems of differing levels. In the meantime, hierarchy theory at least provides a conceptual model to aid in "thinking like a mountain."

Management to protect complexity requires management to protect the *wild processes* of nature, even while recognizing that the best we can do represents a patchy landscape with natural and intensely used areas intermixed. Wilderness areas, which act as repositories for highly evolved and wide-ranging species that exist nowhere else, are therefore essential elements of total diversity, which must be embodied in the patchy landscape that is the American continent in the future.

This management approach accepts the fact that Native American cultures altered the landscape significantly, but distinguishes between those activities and those of plains wheat farmers in the 1920s because the latter, and not the former, disturbed the stable functioning of the larger land system. Whether justified metaphysically, or as a tool for understanding, the organicist image guides the environmentalists' search for a policy consensus on the protection of biological resources.

Environmentalists' emphases on scale, on systematicity, and on process are linked to their preference for the organic metaphor, and some environmentalists emphasize the mystical aspects of organicism. Without reliance on mysticism, however, contextualism, and a recognition of limits inherent in the larger systems environing our culture, can guide steps to protect diversity and complexity. Nonetheless, the theological roots of the search for ultimate explanations of creativity still exist in the ecstatic conception of science espoused by modern environmentalists. While they have not agreed upon a worldview to express the respect they feel for the biological creativity of complex natural processes, environmentalists have agreed that, for scientific, prudential, *and* moral reasons, there must be limits to the damage that human cultures do to living systems.

The Aggregators and Moralists Revisited

Let us return for a moment to the stage at the Smithsonian, where I felt the environmentalists' dilemma so acutely. The schizophrenia that afflicted me on the sand dollar beach was brilliantly articulated for me there on the stage. My conservationist persona was ably represented, right there on the stage of the Museum of Natural History, by the resource economists Hanemann and Randall; the moralist in me was there exemplified by the respected ecologist Ehrenfeld. So how does the philosopher decide? Should I go with the Moralists or the economic Aggregators? This is where philosophers should (if they ever do) earn their keep. It was my chance to be Solomon for fifteen minutes.

In the end I had no Solomonic solution. There could be none. Whereas Solomon faced the dispute between two mothers with the assurance that only one could be the mother of the child, I doubt that the environmentalists' dilemma can be resolved by choosing either Aggregationism or Moralism. This much is clear, however: Aggregationism and the reductionistic language of economics is ill suited to values so complex as those environmentalists pursue in protecting biological complexity. The language of static individual preferences can never express environmentalists' concern for values that depend on the stable natural context that shaped those cultural values nor can it comprehend the value of shifts of perception—transformations such as Muir's elaboration of pantheism after an experience with orchids in the wilderness.

But I am also disinclined to go so far as Ehrenfeld, who sees all economic quantification as obfuscatory, and who attributes rights to all species. To say that other species have rights is to attribute to them a characteristic applicable to human individuals and to assume a competition between human interests and the interests of nonhuman species. Down this road lies either dogmatism or paralysis.[27]

Fortunately, there is another alternative to simple Aggregationism. Not all economists are Aggregators. There exists a strong tradition in resource analysis that recognizes the limits of quantification and purely aggregative approaches to decision making in environmental contexts. The classic book in this minority tradition, which an economist friend once described to me as "one of those books that everyone cites, but nobody reads," is S. V. Ciriacy-Wantrup's *Resource Conservation: Economics and Politics*.[28] Ciriacy-Wantrup advocated a decision criterion that he called "the Safe Minimum Standard of Conservation" (SMS). The SMS criterion, put simply, recommends that a resource be preserved unless the costs of doing this are prohibitive. It is virtually identical to the criterion now implied by environmentalists' policy of placing a heavy burden of proof on any interests threatening to cause an extinction. The SMS criterion operationalizes this standard by declining to quantify benefits of preservation—it assumes preservation is desirable if possible—but quantifies the costs of protection efforts. It sets as its goal to save all species, but accepts that efforts to save species must be politically and ecologically viable, and that choices will have to be made as to how preservation dollars will be spent. The SMS criterion states the commonsense position: In the extreme case, costs might override the strong presumption in favor of preservation, but the burden of proof always rests on those who would degrade a resource or destroy a species.

Richard Bishop, a resource economist at the University of Wisconsin, has applied this concept to a number of specific cases of endangered species, and has concluded that many preservation efforts will not be very costly. This tradition in resource economics, hardly noted in the melee between the Aggregators and the Moralists, has been very important in the evolution of endangered species policy because Oliver Ray Stanton, who served as Chief Staff Economist in the Endangered Species Office during its formative years, is a strong advocate of the SMS approach. Stanton's position, which was not always supported by his superiors at Interior (especially under Reagan and James Watt), was that biological criteria should be kept separate from, and given precedence over, economic criteria in preservation decisions. Practically, for example, this means that Stanton believed that a species should be *listed as endangered* prior to any designation of critical habitat, which required an economic analysis.

In Reagan's Department of Interior, listing proposals were stalled by Reagan's Solicitor General's office, which refused virtually all economic analyses of critical habitat designations as insufficiently detailed. With the support of environmentalists and professionals in the Endangered Species Office, Congress amended the Endangered Species Act in 1984 to reflect this separation and instituted a two-step review process that separated biological from economic considerations. The first step is primarily biological, and assesses the likelihood of extinction, chances of recovery, and so on. At this stage, it is assumed that all species will be protected; laws against the taking of a listed species, for example, would go into effect based on this first determination. At a second stage, economic evaluation determines which means, types of methods, and so on will be chosen for preservation and recovery.

The SMS position differs only in degree from that of Aggregators, provided Aggregators recognize a presumption in favor of the preservation option. The strength of this presumption—stretching from a tendency to settle all close calls in favor of the preservation option to Ehrenfeld's strong claim that all species have a nonnegotiable right to continued existence—

represents a continuum of commitment to environmental as opposed to economic values.

While environmentalists disagree regarding how literally to take organicism, they agree on a central point that is sufficient to define a coherent alternative to a policy based only on Aggregationism: a properly contextualist approach to environmental policy must conceive its object *systematically,* not atomistically and reductionistically. Systems thinking sets environmentalists apart from simple Aggregators. The emerging consensus regarding protection of biological diversity embodies the idea of systems theory as a means to express the common denominator goal of protecting the complexity of the environmental context, to avoid destruction of energy pathways, and to protect the *total* diversity of landscapes.

Meanwhile, as environmentalists evolved toward a consensus on biodiversity policy, the question of standing has been resolved, largely in the favor of environmentalists. Standing requirements have been weakened without the introduction of rights of nonhumans into legal discourse. The trend began in *Sierra Club* v. *Morton* (even though environmentalists lost the case) and continued in subsequent cases. The new criterion for standing has provided environmentalists with access to the court system,[29] and the controversy over rights is now discussed mainly in moral philosophy, not in legal theory.

Appeals to rights of nonhuman species proved judicially unnecessary and politically unacceptable. And yet environmentalists have been remarkably successful in the courts and maintain a congressional coalition that often blocks attempts to reduce important environmental decisions to economic ones. Environmentalists have tried, with varied degrees of success, to resist atomistic models and simple Aggregationism by insisting on a systems approach that values a species in its larger context, and assumes that all species will be saved if the costs are bearable. The contextual approach shifts the burden of proof to those who would cause rapid and irreversible changes in the biotic system. The emerging consensus among environmentalists regarding biological complexity therefore rests not on nonanthropocentrism, but on a growing recognition of the systematic nature of our biological context and an associated realization that the good life must be an ecologically informed life.

9

Land Use Policy

Natural, By Degrees

Albert Hochbaum, whom we met in Chapter 3, was Leopold's student and friend; Director of the Delta Duck Station in Manitoba, Canada; and a part-time collaborator on *A Sand County Almanac*. He also had an admirable talent for succinctly hitting the nail on the head. He summed up Leopold's message in four words. "The lesson you wish to put across is the lesson that must be taught," he said, "preservation of the natural."[1] So much for succinctness; the difficult problem, of course, is to explain what is meant by "preservation" and by "natural."

Thomas McNamee, writing forty years later, uses the same basic approach: "I believe that the true object of conservation is *nature*," he says. "What is nature?" The answer cannot help but be complicated, he notes, because "our conception of nature springs from the darkest depths of our culture's unconscious sense of life itself, and ancient irrational urges and fears give the concept its power.'" But that is only half of the story: "At the same time," he says, "nature must also have an objective, rational, manageable, thinkable value."[2] And thus we have the paradox of modern land use theory: Americans love nature; our values were formed in nature's womb, a huge, wonderful, and horrible wild place. Our values are freedom and independence, "split rail values," as Leopold called them. But our activities, as builders and consumers, transform our environment into something not-wild; we manipulate and control and artificialize nature; we make it not-nature. As the song says, you always hurt the one you love.

But the paradox has also an optimistic face: As we have built and consumed, we have become wealthy by exploiting nature. Wildness has become valuable, objectively, according even to economists, because our wealthy society is now willing to pay to preserve nature. But here is the bitter pill to swallow: We all must admit that, at least in some sense, "nature" preservation is a sham—we've gone too far to "free" nature, as we might free a wild animal, release it from captivity. Just as an animal in captivity loses its essential skills for living wild, our landscape—the land— has lost its ability to equilibrate impacts of human activities.

The challenge of contemporary environmentalism is to develop a response to this paradox, a meaningful guide to land management, one that accepts that we cannot "go back" to a time when nature functioned independently of human impacts. If we accept the environmentalists' dilemma, and see ourselves at a crossroads, both roads look unattractive. If we capitulate to the demands of the land market, in which we all know nature fares ill, we accept a future with more and more artificiality. Down that road lies a Sisyphean task—to find a technological solution to every problem our demand-driven economy creates. Or will we turn toward wildness, reject modern industrial society and its values, and choose nature and wildness for its own sake, even at the expense of human needs? Will we blow up the dams, loose the domestic animals, and go back to the state of nature?

If we are to break a new path, we must set our sights between artificiality—the choice to extinguish nature, to control it everywhere—and primitivism—the choice to isolate and save nature for its own sake; we must escape the environmentalists' dilemma by creating a culture that values nature independently of human demands, but not independently of culture itself. If there is to be a middle path, we must give up the idea that something must be either natural *or* artificial. Naturalness (wildness, too) must admit of degrees.

McNamee illustrates more-or-less nature with examples: A raptor dives for a pigeon at the birdfeeder in his brick-walled back yard in Manhattan. This, McNamee guesses, is "the rock-bottom limit," one of the most compromised cases we'll still call natural, a borderline case. Picnic grounds, national forests where "nature" functions as a wood factory, multiple-use forests, wilderness area—these are all points on the continuum. At the other end of it, opposite to the raptor in a Manhattan yard, McNamee places our greatest national parks. These "encompass vast tracts of magnificent wild country, and in them alone is it possible to make the conservation of nature the dominant value."[3] And, while "natural" must admittedly be a matter of degree, McNamee incisively defines its central variable: "What seems to me essential in the idea of nature as the object of conservation is *wildness*—wildness not just of individual organisms but of places, situations, processes, ecosystems."[4]

Wildness in this very broad sense just means "that which is unaltered by humans"; environmentalists understand wildness to mean autonomy of the larger, environing system. Value is placed on the integrity and autonomy of systems that function independently of human control. In this sense, environmentalists advocate "managing for wildness," or "naturalness," by degrees. They hope to retain and restore the original, hierarchical complexity—the integrity—of environing systems, so that these can function autonomously.

This more-or-less account of wilderness explains, for example, the attitude of environmentalists to wolf reintroductions in remote areas. Environmentalists recognize that top-level predators are an essential part of creative systems. These species close the competitive cycle by regulating populations of grazers, forcing them to adapt not only to available vegetation but also to threats of predation. They create a limiting context within which grazers compete, and hence represent the constraint that must accompany "freedom," the creativity of species to explore new strategies of survival. To understand the ultimate questions of biology, environmentalists must understand how hierarchical systems can generate new life from death, how nature can be creative.

Wolf reintroductions, while potentially difficult, are attractive to environmentalists because they represent respect for the creative processes of natural systems. Those systems, after all, represent also the context of human creativity, the constraints within which cultures struggle to carve out niches. Environmentalists place a high priority on protecting biological complexity, the processes and dynamic systems that embody diversity of natural contexts across time. As a means to illustrate how environmentalists, with Leopold's land ethic as a guide, have struck out into the wilderness between the two roads, searching for an alternative to atomistic economic management and isolationist, antihuman management, we will examine three proposals for land use management of large areas of the United States. First, we will examine environmentalists' plan to manage the Greater Yellowstone area as a "whole" ecosystem; second, we will review environmentalists' proposal that the forests of the Pacific Northwest be managed to integrate forestry and broader goals of protecting biological diversity; and third, we will examine calls for the Chesapeake Bay watershed to be managed as a complex system. These three examples occupy points on McNamee's continuum, and they may help us to see that environmental managers are learning the way between the horns of the environmentalists' dilemma in the area of land use.

THE YELLOWSTONE COMPLEX

In the winter of 1961, 4300 Yellowstone elk were rounded up and shot by Park Service employees. Park management had attempted to prepare the public for the necessary carnage, enlisting independent scientists to corroborate the need for a large herd reduction.[5] Since natural predators had been removed in the 1910s and 1920s, the herd had expanded at the expense of smaller grazers such as antelope and bighorn sheep. New evidence suggested that overgrazing of aspen and willow had caused a decline in beaver, whose abandoned ponds were an essential part of the wildlife cycle at Yellowstone.[6] Live trapping and removal of elk was no longer possible: management could place no more live elk with zoos or wildlife sanctuaries. Controlled hunts, while popular with hunters, had been unpopular with preservationists. Popular or unpopular, it was clear they were unsuccessful in keeping up with the population explosion of elk. Between 1935 and 1961, 58,000 elk had been removed from the Park, but the elk population remained too large, and the quality of the winter browse had steadily deteriorated to a critical level.[7]

Despite attempts to prepare the public and careful planning to keep the slaughter out of the public eye, the outcry was overwhelming. The Park Service was once again put on the defensive.

The National Parks Act of 1916 had charged the new agency "to conserve the scenery and the natural and historic objects and the wildlife therein, and to provide for the enjoyment of the same in such a manner and by such means as will leave them unimpaired for future generations."[8] This dual charge, to conserve the parks "unimpaired" and to make them available for public enjoyment eventually created the ongoing conflict over park management policies.[9] But the immediate problem of the first director of the Park Service, Steven Mather, and his assistant, Horace Albright, was to establish the independence of the new service in the face of attempts by the Forest Service to annex it.[10] They accordingly devoted their early efforts to the development of an independent political constituency.

Both Mather and Albright were excellent promoters and, by organizing sumptuous camping trips for members of the society's power elite, built impressive support for their fledgling agency. Following Pinchot's example of currying the support of the timber industry, Mather and Albright catered to the railroads and other businessmen who could profit from tourism. Albright, for example, had a stock speech entitled "Parks are Good Business," which he delivered whenever possible to Chambers of Commerce.[11] Despite their notable success in promotion, Mather and Albright failed to bequeath a philosophy of management that would keep the dual charges of their new bureau in reasonable balance.[12]

In the early years of park management, no real conflict was perceived between the two goals of the park system. The preservationists, including Muir, saw the parks as one more aspect of progress—their protection as "pleasuring grounds" was considered a useful and important complement to industrializing society, not as directly opposed to it.[13] Furthermore, Muir and his followers saw visitations to the parks by the public as their best means to build a more appreciative attitude toward wild places.

At Yellowstone, this easy attitude toward the twin goals of preservation and tourism led the Park Service to manipulate populations and alter systems to please visitors. Convinced that elk and antelope were the big draw, Park Service managers increased their numbers in the 1910s and 1920s by winter feeding and predator control.[14] Management, meanwhile, vacillated between boosterism, which was successful in building Park Service power in Congress, and occasional unpopular (and short-lived) attempts to apply scientific management principles.

In response to the 1961 crisis, Stewart Udall, President Kennedy's new Secretary of the Interior, convened a commission, to be chaired by A. Starker Leopold, son of Aldo Leopold and a respected Professor of Zoology at the University of California.[15] The result of the committee's deliberations, now referred to as the "Leopold Report," were presented to Secretary Udall in March 1963. It supported the reductions in the elk herd and strongly endorsed the use of science in support of management goals. But the report also suggested a redefinition of the Park Service's management philosophy. The Leopold Report argued that the national parks should fulfill a historical purpose: "As a primary goal, we would recommend that the biotic associations within each park be maintained, or where necessary re-created, as nearly as possible in the condition that prevailed when the area was first visited by the white man." They recommended that "observable artificiality in any form must be minimized and obscured in every possible way." The parks should be managed, according to the report, so that "a reasonable illusion of primitive America could be re-created, using the utmost in skill, judgment and ecological sensitivity." The report therefore recommended an increase in historical research in order to determine conditions prior to settlement by whites, and ecological research on how to recreate and maintain these conditions. Paradoxically, the report set this goal and also acknowledged that achievement of the goal would require "a set of ecologic skills unknown in the country today."[16]

Udall was at first reluctant to endorse the committee's recommendations. But the enthusiastic support of the environmental community for the report convinced him to implement it as Park Service policy.[17] Looking back, it is easy to ridicule the report as naive, recognizing as it did that primitive America could never be recovered and yet urging that the Park Service re-create "vignettes" of days gone by. The report evokes images of Park Service employees as invisible puppeteers, pulling strings from behind false facades of nature, duping the American public into enjoying the impossible dream of time travel. Chase, the most acerbic of recent critics of Park Service policy, says: "The Leopold Report had, in short, inadvertently replaced science with nostalgia, subverting the goal it had set out to support."[18]

But the Leopold report was nonetheless influential. It did for park management problems what Rachel Carson did for pollution problems; it ecologized them. It has, however, taken more than two decades to decide what an ecologized management plan for Yellowstone should look like. Committee members, including Leopold himself, later regretted the "vignette" language, because it suggested a freeze-frame style of preservation.[19] The report appealed to environmentalists because it emphasized the importance of active park management according to scientific principles, with ecological science—not economics—as its guiding scientific light. For park managers, the report provided what they had so long lacked—at least some general principles for managing the natural systems entrusted to their care. The fact is that the Leopold Report, however flawed, represented the first serious attempt to develop a coherent plan for management of visitors and the communities of plants and animals in the park under unified principles. Since acceptance of the report, a consensus has developed that discussion should focus not on static vignettes, but on restoring and maintaining natural processes, or "moving pictures."[20]

Environmentalists and a cadre of activist scientists strongly advocate "whole ecosystem management." By unifying the management of the entire Greater Yellowstone Ecosystem, including the seven surrounding national forests as well as Yellowstone and Grand Teton National Park, environmentalists believe that the area can be managed as a nearly natural system.[21] The argument, as MacNamee puts it, is that the national parks, can be "treasuries not only of the forces that inspire human awe but also of biological processes free of human influence. That is, they can be cathedrals and laboratories at the same time." This, to MacNamee, suggests a general conception of management that could attain social and political support: "to whatever extent the naturalness of a wild place is degraded, both the spiritual experience of it and the scientific experience inevitably suffer."[22]

McNamee does not mean, of course, to return to any primitive vignette, a snapshot of the system Native Americans had negotiated with the land just before the European settlers arrived. He means rather that the management *ideal* should be one of protecting the autonomous processes that were in place at that time, recognizing that rapidly accelerated change has occurred because of heavy extractive and recreational use in only two centuries. It is a plea to protect "the indigenous diversity of the [area] in the largest context, to ensure not just species richness (number of species) but structural and successional diversity across the landscape," in the words of Johnson and Agee. They note that "the scale of the landscape is larger than any traditonally dealt with in natural resources management." Nevertheless, environmentalists have agreed that this is a worthy, even essential social goal.[23]

Whole ecosystem management of the Greater Yellowstone complex, then, has become a central goal with extremely wide support among environmentalists; has received enthusiastic support from Park Service personnel, especially those with scientific training; and has considerable support in the Forest Service. This common-denominator objective, which emphasizes protecting autonomous processes in the largest national park in the lower forty-eight states, has become a rallying point for environmentalists.

But there have been serious misunderstandings. Some critics, such as Chase, virtually identify whole ecosystem management with another concept, "natural regulation."[24] It is very important to keep these ideas separate. Natural regulation refers to a theory that removal of human interference from natural processes is managerially beneficial *wherever possible*. It sets up natural management as an ideal and recommends experiments in restoration—with restoration being understood as restoring a process, an energy flow that existed recently, but was rapidly destroyed by human activities. Ecosystem management, on the other hand, represents a general theory that management problems should be understood in their larger ecological context; it offers an alternative to atomistic management principles. Natural regulation advocates a sliding scale of management strategies, based on the general rule that protecting autonomy is preferable to interference, other things being equal. Natural regulation is so controversial in the Yellowstone area because a growing consensus supports truly managing Yellowstone for the purpose of maximizing its wild processes. As the only region in the forty-eight contiguous states where natural processes are sufficiently intact to allow us to strive to actually achieve that ideal, the Yellowstone area is perceived by ecologists and environmentalists as a last chance to maintain a place of wildness for scientific enlightenment and spiritual inspiration.

Neither ecosystem management nor natural regulation, if properly understood, implies walking away from a volatile situation where human impacts have greatly degraded the autonomy and complexity of the system. The crying need—if park managers are going to treat Yellowstone as a wild ecosystem requiring gradual development and restoration so that eventually a new equilibrium, mainly uncontrolled by managers, would emerge—is scientific work. Park managers are not wrong in trying to implement natural regulation; they are wrong in implementing it without proper attention to ecological and historical research. For example, park managers and the scientists who work for them have been criticized for applying natural regulation to the northern range of Yellowstone, where it has been assumed for decades that bison, and especially elk, had caused range deterioration. Managers and scientists cite photographs, a few going back over a century, showing that the northern range was even then less lush than other Yellowstone ranges. Critics insist this evidence is too slim to declare out of hand that overpopulation of elk and bison no longer exists.[25]

A proper policy of natural regulation is not passive; it actively seeks ecological and historical data, trying to determine the autonomous trajectory of natural systems, and then institutes cautious experiments in restoration. The most important fault of the Park Service has been its failure to develop an adequate science program and to utilize science properly in management planning. On this point, there is more than enough blame to go around. One could blame Park Service bureaucrats who are more interested in managing people than natural systems; one could blame the service's structure, which virtually ensures that park supervisors will come up through the ranks as people managers rather than managers of nature; one could blame Congress, which anxiously supports road-building and development in the parks, but which provides embarrassingly small allocations for scientific research; or one can blame the American people, who insist on roads and hotels, but not science. Every commission and every study group that has ever looked seriously at Park Service policy has concluded that the present situation is intolerable. The mandate of the Park Service to preserve and protect the parks as natural areas will be meaningless until the service, with support from Congress, institutes a strong and independent science program and devises means by which scientific studies can legitimately guide park managers toward their goals of "preserving wildness."

It is important, however, to understand these debates against a backdrop of agreement. Both park managers and their critics agree that the goal should be to protect, to the extent possible, the natural processes that constitute the magnificent natural systems that exist in the greater Yellowstone ecosystem. Disagreements over scientific assessments and whether a given application of the policy of natural regulation is justified must be understood in the context of this common-denominator objective.

OLD-GROWTH FORESTS OF THE PACIFIC NORTHWEST

Timber interests disparagingly refer to the spotted owl as "the billion-dollar bird." The populations of this shy and retiring bird have been declining steadily, and the species has become the center of a pitched battle between environmentalists and lumber producers, with Forest Service and Bureau of Land Management (BLM) lands as the battlefield. The northern spotted owl depends on old-growth forests of the Pacific Northwest for its survival. These forests have declined preciptiously under the lumberjack's saw and now cover only one-fifth of their original 15 million acres, with most of the remainder existing in national forest and BLM land. The economically pressed forest industry insists that it needs to log the remaining old-growth forest to stay in business, and that the spotted owl is too costly to save. The industry has bitterly opposed changes in Forest Service leasing policies— based on a scientific commission and reluctantly supported by President Bush—which are designed to protect the spotted owl, arguing that they will be economically disastrous.

Environmentalists, however, respond that placing a price tag on a predator species such as the spotted owl, as if it were a parakeet in a pet-shop window, may result from asking the wrong questions. The forests of the Pacific Northwest are among the oldest in America and the spotted owl has been integrated by millennia of competion and evolution into the multi-storied habitat created under the canopy of 200-foot-high Douglas fir and hemlock; its fate cannot be separated from that of the old-growth forest as a working ecosystem. A breeding pair of spotted owls requires anywhere from 2000 to 4000 acres of old-growth forest, depending on location. As the forests on which they depend are fragmented, the owls are only the first of many species to be threatened—for the Pacific Northwest forests contain perhaps the largest number of bird species of any forests in the United States, together with such other creatures as bats, flying squirrels, and martens. These species depend, in turn, on a rich plant and invertebrate population that is generated from decaying logs on the forest floor. The timber barons call this "waste" and refer to mature forests as "cellulose cemeteries," but it is in these logs that the circle of dependencies in old-growth forests is completed: The trees on which the animals depend each depend, in turn, on the animals—red-backed voles eat the fruiting bodies of fungi on dead logs and excrete their spores on new sites. These spores grow into fungi that carry necessary nutrients and moisture to tree seedlings.

Specialized species such as the spotted owl represent the vanguard of evolution and, because they are so integrated into the old-growth systems on which they depend, their protection cannot be undertaken without protecting significant patches of the systems themselves. Protecting the spotted owl therefore becomes a symbol for protecting an entire bioregion; and, looked at from the side of everyday management rather than from the side of goal-setting, the spotted owl is also important because its populations serve as an indicator of how we are doing in protecting that bioregion on a larger scale. The loss of the spotted owl would signal the demise of a complex web of relationships, and the fragmentation that causes such a demise will lead in turn to the demise of an irreplaceable, interconnected system.

Thus, by examining the interaction of social goals, including the goal of protecting total diversity of bioregions, and the emerging scientific evidence about how many acres of old-growth are necessary to maintain safe populations of spotted owls, we can begin to form hypotheses about how to save the owls *and* the systems they depend upon. Two common-denominator objectives emerge from this analysis: (1) to protect total diversity of the Pacific Northwest Forest as indicated by its ability to maintain one of its "pinnacle species," and (2) to use public lands to design a landscape that represents a viable compromise between the exploitation that will unavoidably take place on private lands and at the same time allow for the constraints imposed by (1). Whole ecosystem management, here practiced over an even larger area, with even more complex mixes of productive, ecological, and aesthetic goals, is essential to achieve, or approximate, both of these common-denominator objectives.

The issue, of course, is how much old growth to save and how it should be configured to retain the total diversity and general character of the bioregion. Implicit in this question is the implication that, if the requirements for protection are too high, and the costs in terms of forgone opportunities to generate timber too great, we may have to sacrifice the spotted owl. Political realities, once again, become a part of the management equation. It is on these interlocked issues that the environmentalists do battle with the exploitationists, with the Forest Service often caught uncomfortably in the middle. They are under tremendous political pressure to use the national forests to maximize production while they are charged by Congress to protect biodiversity on Forest Service lands. They experience, in a concrete way, the environmentalists' dilemma.

In a remarkable and not uncontroversial book, Larry Harris has proposed a hypothesis—a boldly theoretical guide to management decisions—based on the theory of island biogeography, the study of patterns in biotic communities on oceanic islands. Developed in the late 1960s, island biogeography has a long history, predating even the voyages of Darwin and Wallace, but information, principles, and theory have been developed into a cohesive theoretical framework only over the past thirty years.[26] Islands provide useful subunits within which to study the dynamics of populations, and the special vulnerability of island species to extirpation makes them especially interesting to conservation biologists. Noting that preserves designed to protect biotic diversity share structural similarities with oceanic islands, Harris says, "If the biological principles and generalities from true islands are applicable to patches of old growth in the forest landscape, the biological foundations of a planning strategy can be established."[27]

Harris's management plan, which he calls "the island archipelago approach," involves shifting emphasis "from the old-growth system to the system of old growth."[28] That is, the plan focuses not on the content but on the context of the old-growth islands. Harris says, "The only level of the hierarchy [of biological communities] that is both necessary and sufficient to meet all objectives is the ecosystem or some higher-level approach. The strategy selected should not only ensure the conservation of spotted owls, but all the intricate linkages that are associated with natural populations of spotted owls in naturally functioning ecosystems. Many of these are as yet unknown."[29] As Leopold realized, management is always experimental.

Island biogeography tells us that, other things being equal, a small island will have proportionately fewer species than a large island; it also tells us that species diversity will vary inversely with the distance isolating the island from a continent that can serve as a source of colonizing species. One important absence of analogy between oceanic islands and nature preserves within a landscape is that, in cases of relatively developed landscapes, there exists no "continent"—no vast repository of ecosystems and species fulfills this function. The system must then be designed to be self-sufficient. Fortunately, in the Pacific Northwest there is a system of national parks and wilderness areas that are large enough to function as repositories of species, though these are mostly at elevations too high to have great species diversity. Despite their inadequacies, these areas provide the "building blocks upon which a conservation strategy may be hinged."[30]

Harris's approach to protecting ecosystem-wide diversity is to design site-specific protection areas so they function as an integrated landscape system. Central to such a system is protection of old-growth islands. Political and economic constraints will limit the size of the islands, so Harris designs systems with long-rotation islands surrounding islands of protected old growth. Each of these islands is composed, in addition to its old-growth core, of nine even-aged stands cut in an alternating sequence designed to maximize edge effects between stands. Cutting one of the stands every 35 years results in a 315-year cycle as represented in Figure 9.1.

These long-rotation stands serve as a buffer around the old-growth stands, protecting them from the effects of the developed landscape, as well as from agents such as wind, increasing the effective size of old-growth stands as contributors to total diversity. This long-rotation system is designed to allow significant timber productivity while protecting diversity by counteracting the effect of diminishing diversity on smaller islands.

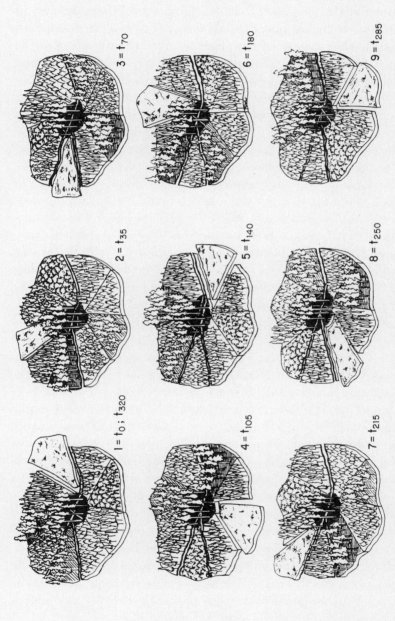

Figure 9.1. An old-growth patch surrounded by a long-rotation island that is cut in a programmed sequence (alternate stands) that leads to maximum average age difference between adjacent stands over a complete cutting cycle. (From Larry D. Harris, *The Fragmented Forest: Island Biogeography Theory and the Preservation of Biotic Diversity* [Chicago: University of Chicago Press, 1984]. Used by permission.)

The first threats to the total diversity of an area occur when wide-ranging species with large habitat requirements, such as raptors and larger carnivores, are lost because of habitat fragmentation.[31] These species will often have much larger ranges than can be accommodated within any particular habitat island. The "effective island size"—its true worth as a protector of these wide-ranging species—will be determined by three factors: the actual size of the island, the distance between islands, and the hostility of the matrix between islands.[32] Harris suggests that protection per unit space can be maximized if the mosaic is designed to provide "habitat corridors" among islands. These corridors, sometimes quite narrow strips of protected or semiprotected habitat, encourage movement among islands, reduce genetic isolation of populations, and provide wide-ranging species with access to more than one protected area and therefore an increased range of protection. Harris suggests that, wherever possible, strips along rivers and streams be protected for this purpose.[33]

Harris's proposal, then, is to use the lands of the Forest Service and of the Bureau of Land Management as the backbone of a landscape-wide plan to protect the biological diversity of the Cascade mountain range in Washington, Oregon, and Northern California. While Harris's plan is controversial in some of its details, and will become increasingly so when it is applied to varied particular situations, the island archipelago approach does illustrate an emerging consensus among environmentalists in its emphasis on contextualism in management.

CHESAPEAKE BAY

Calls to "Save the Bay" are now more common in the Chesapeake region than rockfish; a collective fear is beginning, only slowly in some quarters, to grip the entire area: "The bay is dying."

Strictly speaking, the bay will not "die" in the sense that all life will be extinguished there. The larger bay system, however, is in grave danger of shifting to a new and less desirable level of functioning due to a rapid change in the ecological balances and a serious reduction in complexity of the system. Populations of phytoplankton, naturally occurring communities of microscopic plant cells and algae, are exploding as a result of estuary-wide increases in nutrients, especially nitrogen and phosphorous, that are pouring into the bay from sewerage discharges and from agricultural and suburban runoffs. The rise in phytoplankton populations reduces water clarity, resulting in a severe depletion of the underwater grasses that have traditionally absorbed these nutrients and passed them up the food chain to larger animals. Canvasback ducks, for example, which feed on the grasses, have nearly disappeared from the area.

There are other signs of intense stress in the magnificent bay ecosystem:

heavily saline water that collects in the deep areas of the bay remains oxygen-poor for longer and longer periods each summer; harvests of all finfish and shellfish, except the hearty blue crab, have dropped steadily. Taking of rockfish, a symbol of the bay's bounty to commercial and sport fishermen, was halted in 1985 because populations were badly out of balance and declining rapidly. A brief, heavily regulated season was re-instituted in 1990.

While the bay is not dying, the bay *is* sick; its essential processes are being interrupted and altered by human-induced change. The pathways through the large, normally slow-changing and complex system that pushes productivity upward through the biotic pyramid from algae to ducks, eagles, and humans, is being clogged near the bottom by human wastes. Management of the Chesapeake region represents an extreme test of the general goal of conserving nature; the bay represents an example of a system brought nearly to its knees. Here we must ask what would *count* as "preservation of the natural" in the heavily developed and still rapidly developing bay region? One point is clear: It is impossible to turn the clock back to an earlier, predevelopment stage in the process. Relinquishing control of bay systems would at this point be disastrous, even if possible.

Spurred by a major water quality study completed by the Environmental Protection Agency in 1983, a remarkable political consensus has emerged in the Chesapeake region. A bay summit was called in December of 1983, and politicians from Maryland, Virginia, Pennsylvania, and the District of Columbia united in a compact pledging cooperation and resources to a regionwide effort to clean up the bay.[34] An Executive Council, with an Implementation Committee, Citizens Advisory Committee, and a Scientific and Technical Advisory Committee has been instituted and participants have accepted a goal of reducing nutrient runoff by 40 percent by the year 2000.[35] The U.S. Environmental Protection Agency, which set up its Chesapeake Bay Program in 1976, has fully cooperated with the states and has embraced with them a holistic, organicist conception of bay health and the clean-up problem. Deterioration of bay quality, all participants seem to agree, can be reversed only by a cooperative effort to go beyond atomistic thinking to direct attention toward the overall health of the entire bay system.

Operationally, thinking organically about the bay entails expanding the *temporal* and *spatial* context of bay management. "We are throwing out our old maps of the bay," says Tom Horton, a popular writer on bay culture and manager of an environmental education center on Smith Island, in the middle of the Chesapeake. "They are outdated not because of shoaling, or erosion or political boundary shifts, but because the public needs a radically new perception of North America's greatest estuary." The new maps include all of the bay's tributaries and, on them, the bay proper "is a relatively small puddle, surrounded—almost overwhelmed—by the lands of a watershed 30 times its size."[36] Thinking of the bay as a living system also requires a recognition that it is a dynamic system, changing and adapting through time.[37]

The causes of bay deterioration, it is now clear to everyone involved, cannot be traced solely to some specific activity of human polluters, but simply to population growth itself, and to functions that arrive naturally with that growth. It took 350 years for human population in the watershed to reach eight million. Then, in the past thirty years, the population shot up another 50 percent and it continues to grow, especially in the ecologically sensitive waterfront areas. So the Chesapeake Bay region is at the cutting edge of environmental problems and the search for solutions. As Horton says, "If humankind were to order up a test case of how many of us can exploit natural systems without irretrievably harming them, we could have constructed the scenario we are living out around the Chesapeake Bay."[38] The U.S. Congress Office of Technology Assessment has recently recommended that the Chesapeake Bay clean-up strategy should be used as a model for addressing similar problems—emerging slightly later—in other estuaries around the country.[39]

Ecosystem management in the Chesapeake Bay region cannot have the same specific goals as management of less heavily used regions such as the Greater Yellowstone Ecosystem or the forest systems of the Pacific Northwest. What is shared in all of these cases, however, is a recognition that management, no matter how intense the use involved, must address environmental problems in their larger context. Political unification behind an experiment in watershedwide preventative and restorative medicine is important, but will succeed only if it is supplemented with rapid advances in scientific understanding and considerable luck. But the costs of failure are not only uncountable, they are in a deeper sense unthinkable.

Conceiving success is unfortunately almost as baffling. However useful natural regulation may be as a guide at Yellowstone, "restoration" of maximal wildness and ecological self-determination would be an inappropriate goal for the Chesapeake Bay region. Habitation of the Chesapeake by European settlers and their offspring has introduced irreversible changes into the bay system; the further a region gets from the natural baseline, the more local history has transformed natural processes, creating a new balance between natural and human forces.

But herein lies the key, if there is one, to integrating humans into their natural environment. The rich Chesapeake Bay culture depends for its meaning on the rich natural context in which it has evolved. Human culture is "natural" to the degree that it retains and protects this integration in the future. As Horton has so richly illustrated in his book, *Bay Country*, saving the unique cultural heritage of the Chesapeake Bay is so intertwined with the task of saving its natural context that they have become a single task. Watermen without oysters will be, at best, cigar-store Indians.

Horton concludes:

> To preserve the full diversity of such connections as [the inexplicable yearning created by Canadian Geese migrating over our cities] into the next century will require a broader view of environmental protection than is likely to evolve through our legal and political systems alone. Carl Jung, the great psychoanalyst, once said he had never been able to cure a patient who did not have a firm belief that he or she was part of something larger. Thus we need our religions, our cosmologies, and equally, I think, a greatly expanded appreciation of all the ways in which we and nature fit together.[40]

We must, in other words, conceive our culture as a part of its natural context, and the natural context as essential to our culture, if we are to conceptualize the shared social goal of "saving the bay." Preserving the natural must, in such cases as the Chesapeake region, be a matter of degree; our guiding light must in these cases be nature as it has evolved within a distinctive culture based in a distinctive natural setting. The goal of management for the Chesapeake is to stabilize the forces that affect the system and reverse the process of deterioration by intensive restoration efforts.

Because changes now perceived were delayed (or at least unnoticeable) for several centuries, we know the bay embodies processes capable of integrating and damping out considerable human impacts. As human development has continued, a new system, incorporating human and natural forces and structures, has evolved. The key to intelligent management in such cases will be to understand human systems as relatively fast-changing systems of agricultural use, changing housing patterns, and so on that impinge upon the larger and slower-changing natural systems that provide the context for human systems. The goal of whole ecosystem management, in the context of a rapidly developing estuarine watershed, means paying constant attention to the contextual effects that result from increments of growth and to environmental stresses caused by changes in economic forces and human taste.

The "system" of hierarchies used to model the context of economic activities must encompass human and nonhuman elements. The adage "preserve the natural" will in such cases mean preserving a mixed artificial and natural system that is sufficiently slow-changing so that new equilibria, new accommodations among human and natural communities, can evolve. Preserving the "natural" means, in these cases, recognizing that many of our activities, if carried out pervasively and persistently over the landscape, may set in motion changes that will destroy the context, and hence the satisfactions, that depend upon the Chesapeake's rich natural and cultural heritage. It means learning to live within the constraints set by the larger natural system that give meaning to local and regional cultures.

Scientifically, hierarchy theory may help to develop useful models to chart interactions between human economic and social systems (which change as a result of trends in decisions made by human individuals) and the larger natural systems that determine bay water clarity and bay productivity. In this sense, the larger bay system envelops many smaller natural systems— the tributaries and marshes that feed the bay—and it is these natural inter- actions that must be understood and monitored as they play themselves out in shifting patterns of submerged grasses, algal blooms, and so on.

Whatever else it may mean, the decision to think about bay health en- tails developing a detailed model of how comparatively rapid trends in hu- man activities will affect bay processes. These processes will continue to change dynamically, but, if the bay is to regain its health, causal influences from rapid changes in human activities will have to be slowed, so as to allow the larger, autonomous systems that determine bay quality to absorb new inputs of nutrients, and to develop a new, self-directing system as the larger context of human life around the bay.

Recognizing the importance of protecting the systems that environ the activities of the residents of the Chesapeake Bay area, the Executive Coun- cil is developing "habitat and living-resource objectives which, when com- bined with monitoring data and modeling projections, will tell us how great a reduction of nutrients and other pollutants is required to bring back the Bay."[41] The organization of the council, which incorporates inputs from a Citizens' Advisory Committee, recognizes that holistic and systematic thinking cannot reflect only top-down thinking. On the contrary, a true understand- ing of systematic thinking emphasizes the upward, from the grassroots, thrust of systematic trends. Because the use of natural resources expresses local differences, as local communities and their economies develop ways to use and interact with bay ecosystems, maintenance of a healthy larger context will depend on maintaining distinctiveness and individuality in local land uses.

So far, efforts at bay protection and restoration have emphasized changes in land use planning and attempts to limit especially damaging activities in the immediate environs of bay waters and their tributaries. The center- piece of current efforts in Maryland, for example, has been an ambitious plan to involve local communities in a comprehensive overhaul of local planning. A Critical Areas Commission (CAC), a twenty-five member com- mission representing county and state interests, has set down criteria to guide local governments in writing a plan, which must be approved by the CAC. This approach emphasizes locally originated planning to accommodate growth within extemely broad outlines set by the state legislature—to conserve wildlife habitat, to minimize water quality impacts, and to address problems of development. Every local government must submit for review and approval a land-resource protection program. Lands within one thousand feet of the shoreline of the bay or any of its major tributaries are automatically protected, with maximum densities of one dwelling per twenty acres and encouragement to cluster new development. Recognizing aesthetic and wildlife protection as legitimate objectives of land use planning, the CAC program has broken new ground in encouraging innovative approaches to local planning.[42]

Despite the forward-looking nature of the Maryland approach, however, few observers believe it represents more than a first step toward protecting the bay. The one-thousand-foot rule brings only 1 percent of the land area in the watershed under the central commission's control. Further, provisions allow local discretion to exempt areas that are already 50 percent developed. These provisions suggest that CAC guidelines will have greatest impact on agriculture and development in remote areas. They may hardly affect the rapid development at the fringes of urban areas.[43]

There is also reason to question whether limits on zoning will have any positive effects in the short term. As of 1984, Maryland had 891,000 acres approved for new development, which represents 500 percent of the projected needs of the state for the next 13 years.[44] Worse, a rush of new subdivisions and rezonings was precipitated in anticipation of the stronger controls on growth discussed by the CAC. Local governments have, for the most part, written their plans so as not to interfere with but only to control somewhat the effects of continued development in their areas.

Restrictions on development, although useful in mitigating problems on the immediate edges of the bay and its tributaries, are therefore unlikely, by themselves, to save the bay system. As long as local governments continue to see development as an economic resource, and the bay as a "commons" attracting new second-home customers, it is doubtful that the decline of the Chesapeake will be reversed. The well-meaning and innovative initiatives already begun must be strengthened.

Contextual thinking helps to suggest new directions: The context of locally distinct cultures can be saved only if local communities go beyond negatively conceived restrictions and develop positively understood goals for protecting and developing their indigenous cultures. For rural areas and for local cultures still deeply entwined with local natural resources, this will mean resisting the homogenizing effects of continued growth. Local fisheries will have to be defended against marinas for pleasure boats, farms will have to be protected from encroaching suburban development, and so forth.

These further steps will occur only if local communities take a positive and aggressive approach to defining and defending their traditional, local values. These values must be generated locally and flow upward toward regional planning commissions; they must result from a local determination to protect the unique cultural character and the physical context essential to it. Once positive local values were articulated, planning could move past negative restrictions such as minimum lot sizes and would involve development of incentives to encourage owners to maintain uses that originated indigenously in response to local natural conditions. It would involve finding innovative means to help landowners to choose to keep their land around the bay in less polluting uses such as forest and pasture, rather than to convert them to more polluting uses. Such a positive approach to planning must represent a positive groundswell of local determination to preserve indigenous cultures and, through enabling devices and incentives, to steer trends away from activities that are damaging to the cultural and natural context and toward activities that offset development elsewhere. A positive conception of the *integrity* of the land, of its ecological complexity, and of its creative and productive processes must guide new development. Ecosystem management requires a biologically and ecologically informed criterion to guide future development.

Parts And Wholes

The three applications of ecosystem management and contextual thinking examined here provide a basis for generalizations about the growing consensus, in land use policy, in favor of ecosystem management. First, we can say that whole ecosystem management does *not* entail a mindless commitment to the idea that "nature knows best."[45] While an effort to nudge certain large and unspoiled systems such as the greater Yellowstone ecosystem (GYE) back toward natural regulation represents one legitimate application of ecosystem management, the idea has application also to areas where development has gone too far to permit management to maximize wildness. The essence of ecosystem management, rather, is a commitment to *contextual thinking* in the development of resources. Thus we find Ian McHarg talking about protecting natural processes and local character jointly, as social values, even in developed areas such as Staten Island. McHarg's widely read and cited book, *Design with Nature*, describes a systematic approach to land use planning based on local values and an ecological conception of human habitation as adapted to the natural contours of the land. McHarg emphasizes that planning choices must be sensitive to local ecological and other natural features. Lewis Mumford describes McHarg's proposals in glowing terms:

> Here are the foundations for a civilization that will replace the polluted, buldozed, machine-dominated, dehumanized, explosion-threatened world that is even now disintegrating and disappearing before our eyes. In presenting us with a vision of organic exuberance and human delight, which ecology and ecological design promise to open up for us, McHarg revives the hope for a better world.[46]

The moving force behind this thinking has been a more organicist conception of the human relationship to the land. An organicist conception does not deny human impacts and attempt to go back to an earlier age, but rather recognizes these and attempts to build creatively upon them. Organicism as a guide to management suggests a forward-looking creative management that accepts what has evolved, recognizes human impacts, judges some of those as contributing to health, and seeks to control others as destructive of the integrity of the ecological context. This does not involve treating nature as a nonentity or as a mere source of natural products; it involves treating nature as an ongoing, autonomous system that has absorbed human impacts and has changed in response to them. Neither by denying past impacts nor by ignoring the effects of future impacts, but by humbly and carefully trying to understand and mitigate them, can modern civilizations build a healthy foundation for future growth within a healthy natural context. Contextual, or hierarchical, thinking recognizes that true creativity always emerges within constraints set by a larger system.

With all this said, however, a terrible, perhaps fatal, difficulty remains: Just how seriously can we take any particular designation of a "part" or a "whole" of a management "cell" and its larger "context"? Drawing ecosystem boundaries can be crucial in management decisions—witness the current discussion of whether to expand the Park Service's ecosystem management policy to include the seven national forests surrounding Yellowstone. If environmental management is to be scientific, does it not follow that the management units it focuses upon must be scientifically valid?

A coalition of environmentalists, scientists, and environmental managers have endorsed the idea that Yellowstone and its surroundings should be managed as a whole ecosystem to maximize wildness. Alston Chase has attacked the idea of whole ecosystem management, arguing that it represents "only ritual bows to a mystical sense of connectedness, not serious scientific undertakings to develop a science of ecology."[47] Chase argues that environmentalists, disappointed in the slow development of ecological science, have substituted "California cosmology" for science. By examining Chase's argument and its application to management in the Yellowstone area, we can clarify the idea of ecosystem management and simultaneously explain why the term "contextualism" may be preferable to the more commonly used terms "holism" and "organicism."

Chase's argument begins by reviewing the history of the term "ecosystem," concluding: "The ecosystem, in short, was a tool by which scientists artificially separated their subject of study from everything else. It was an idea, a fiction, for in reality no system was isolated."[48] "The vaunted self-regulating ecosystem, the supreme assumption that guided management of wildlife in Yellowstone," Chase concludes, "was just a fictitious axiom, conceived to satisfy philosophical and mathematical conceptions of symmetry."[49] Leopold's land ethic, Chase argues, which had its goal to integrate human activities into natural systems, was subverted; park ecosystems were treated as functioning independently of humans. "The science of ecology in which Leopold, Carson, and others had such high hopes was falling short of its goal, even as the environmental movement was reaching its stride," according to Chase. The problem, he thinks, is that "while [ecological] scientists moved ever more narrowly into abstract realms of mathematical manipulation, environmentalists traveled more widely in search of a spiritual cosmology of nature."[50]

Chase believes that the idea of whole ecosystem management represents a new form of isolationism—treating systems as "whole," he thinks, is just an excuse for not doing the scientific study and management necessary to integrate human impacts into natural systems. Environmentalists, on his view, use the whole ecosystem myth to cover up failures in Park Service management—failures can always be blamed on the fact that even Yellowstone National Park is not large enough to enclose a whole ecosystem.

In a chapter entitled "The California Cosmologists," Chase argues that environmentalists, who have failed to develop a positive view of man's role in nature and were unable to derive any useful guidance from the fictitious idea of ecosystems, fell back on Eastern religion and various forms of mysticism, seeking guidance rather in new ideologies.[51] While scientists sought technical solutions, "Deep ecologists, mostly humanists and social scientists such as Devall, Rodman, and Sessions, looked to the past and the revival of a humane, nonindustrial ethic." Debate between the scientists and the environmentalists, "and the debate among environmentalists nearly everywhere—became increasingly abstracted, far removed from grassroots problems like the fate of beaver in Yellowstone."[52] Environmentalists sought solutions, "not in better biology, but with the right ideology."[53]

Must advocates of whole ecosystem management choose, as Chase suggests, between scientists' notion of ecosystems as temporary methodological conveniences and mystical holism, a belief in a universe-sized supraorganism that controls and regulates interactions of species from above? To answer this version of the environmentalists' dilemma, it is necessary to note a serious ambiguity in the term "whole," which has at least four mean-ings—two managerial and two philosophical—when it is used in discussions of ecological management. Failure to recognize these ambiguities can lead to serious confusion.

First, and most commonly, there is a *boundary* use of the term, in which "whole ecosystem management" represents no more than a decision to choose an ecologically defined community as the larger, contextual unit defining the limits of the management unit. Whole ecosystem management in this sense recommends, contrary to the usual practice of determining the boundaries of environmental units by political and institutional boundaries, the practice of using ecological concepts to define management units.

There is no single ecological criterion to guide these choices—in the Chesapeake region, the management goal of guarding and improving bay water quality implies management over the whole Chesapeake watershed. In the Yellowstone area, where the goal is managing populations of large mammals, the unit will be chosen as the outer limits of the wanderings and migrations of those populations. When environmentalists use "whole" in this boundary sense, they are opting for a set of boundaries for the opera-tive management unit that will be ecologically defensible, given the stated management goals. This decision should not be confused with the naive view that an area can be "sealed off" from impacts crossing any boundary. Environmentalists recognize that there will be inputs from outside the sys-tem—ecosystem boundaries will always be permeable membranes allowing impacts from as far away as the sun, for example. The choice, rather, is pragmatic: Given the management goal chosen, one must choose a system that is whole enough, ecologically, so that system variables are not over-whelmed by inputs and outputs. The boundary must designate an area that is sufficiently "whole" to represent a manageable unit. Once this decision is made, ecologically, designation of a whole ecosystem amounts merely to drawing a line on a map, thereby designating impacts across that line as either inputs or outputs (see Figure 9.2).

Since the drive toward whole ecosystem management encourages look-ing at very large areas, such as whole watersheds or the range of a healthy population of grizzlies, whole ecosystem management entails management that is sensitive to larger and larger contexts. But whole ecosystem man-agement can be practiced on all scales—management of sewage treatment in Baltimore and Washington, D.C., are very intensive, focused on the goal of reducing damaging outflows from small, intensely used and man-aged areas. Management, nevertheless, can be conceived contextually, and in this case means mainly minimizing spillover impacts of intensely popu-lated areas on their larger natural context. As management looks to larger and larger systems, geographically, hierarchy theory would predict that the

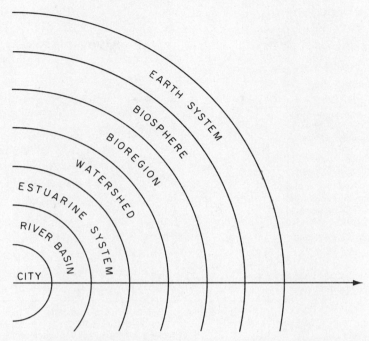

Figure 9.2. Whole Ecosystem Management. Ecosystem boundaries are chosen to be ecologically appropriate, given a chosen management goal.

temporal frame of management goals and strategies must also expand, from daily, practically crisis management in some cities to management for the ages across a larger landscape. This correlation is represented in Figure 9.3.

But "whole" has another managerial sense. When it is proposed that the GYE be managed as a complete ecological system, "whole" is intended in the *biotic* sense: to promote wildness, with the goal of maintaining, and restoring where lost, the multilevel biological *complexity* of the GYE. The goal is to re-create the creative processes that have sustained the complexity of the system through time, to re-create a system that is sufficiently complete to embody all levels of the ecological pyramid. While this purpose requires management over an area large enough to encompass migrations of large herds, it also requires management that is sensitive over very long periods of time. A call to nudge the Yellowstone ecological system back toward wholeness is a call to manage the system to protect its hierarchical complexity and the interlocking processes that create and sustain ecological diversity through time. Management for wildness requires managing a system sufficiently whole in its multilevel complexity to be sustainable through evolutionary as well as ecological time.

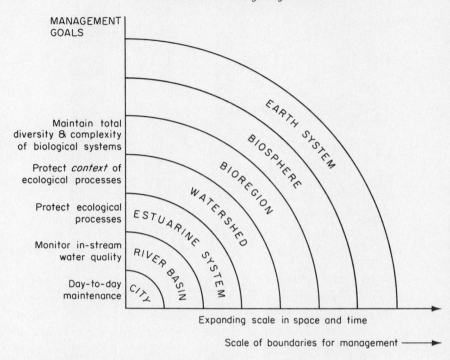

Figure 9.3. The Scale of Whole Ecosystem Management.

A contextual approach to management will focus attention on questions of *scale*, both spatial and temporal, and these two conceptions of whole-ness, the boundary and the hierarchical sense, correspond essentially to these two scales. Contextual management can be successful only if the spa-tial scale of the ecological context is construed broadly enough to encom-pass the major causal factors affecting bay quality (whole watershed). Man-agement to protect the multilevel creativity of nature, as described in the third section of Chapter 7, involves managing in expanding frames of time as well. Thus, while the Chesapeake Bay region is managed as whole in the boundary sense, it seems unlikely that it will be managed as whole in the biotic sense (which would involve attempts at predator re-introduc-tions, for example). In the Chesapeake region, regulation of deer herds will necessarily be "artificial," relying on manipulation of hunting regulations or introducing birth control for deer. The goal is to maintain changes suf-ficiently moderate to offer opportunities for existing species to adapt, *in ecological time*, to their changing context. Management of Yellowstone, on the other hand, aspires to wholeness in both senses, and therefore chooses as its goal to protect the wild ecological *and evolutionary* processes that

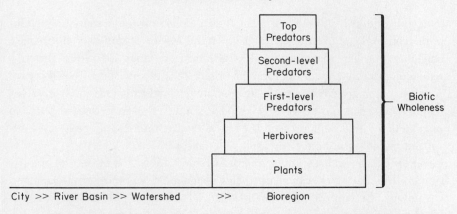

Figure 9.4. Whole Ecosystem Management. Biotic wholeness over a whole biore-gion represents wholeness in a richer sense: the system is managed to protect large-scale ecological and evolutionary processes.

determine the growth and selection of populations *in evolutionary time*. Such management daringly proposes experimenting with initiatives to re-store health to a biotically complete ecological system, one capable of func-tioning autonomously in evolutionary scales of time. We can dramatize the two senses of wholeness in Figure 9.4, which superimposes a whole biotic pyramid on the same continuum as shown in Figure 9.2.

We can now understand more clearly the implication that systems will be managed differently depending on where they fall on the "wildness" continuum, which we can now picture, with a few added points on the continuum, as in Figure 9.3. At the far left of this continuum, cases of large cities and other systems heavily altered by human activities, management will be intensive over a small area and will usually involve a flattened sys-tem with relatively few levels in the biotic pyramid. Management of a river valley may include a large city on its banks, and management of an entire bay watershed may encompass several river valleys with several cities. Management over the larger system will encompass more wild places, and the overall mosaic of land uses must be arranged to protect as much total diversity and complexity as possible, even though no large predatory mam-mals are present in the system. Here, artificial management must adjust populations to levels sustainable by local systems, but attention must also be paid to the overall health of populations that will wax and wane across a dynamically changing landscape. That is, management for protecting complexity must encompass all existing levels, even when the pyramid is truncated, as when predators have been removed.

Only in Yellowstone, of the management areas in the contiguous forty-eight states, can contextual management aspire to "wholeness" in both senses and, consequently, to management that is holistic in both spatial and temporal contexts. Management to protect the entire pyramid of life over a major area implies management through multigenerational, indefinite time—management, that is, in evolutionary time. It is in this sense that the proposal for managing Yellowstone as a whole ecosystem is radical. It involves management not only over a relatively "whole" system in space, but also management that will protect the multilevel complexity of processes resulting in the evolutionary creativity of nature.

When Chase accuses environmentalists and advocates of natural regulation in the Park Service of deserting science in favor of mysticism and cosmology, he confuses the two managerial senses of holism with a metaphysical conception of ecological systems as deified wholes that have value in themselves. But once the boundary sense of holism is clearly distinguished from the biotic sense, we can see that environmentalists *need* not embrace mystical holism to support their whole ecosystem approach or their ideal of natural regulation. The value of protecting a system that is whole in both senses can be justified for many reasons, as MacNamee notes, both scientific and religious. It can be argued on scientific grounds, for example, that the boundaries of management plans must be expanded to cover the whole ranges of an area's migratory animals and that this is implied by the goal of managing a whole ecosystem. As hierarchy theorists argue, the justification for choosing particular matrices of systems is "utilitarian" and "epistemological," not metaphysical. The choice of a whole system to manage should be viewed as an aid to understanding, not as inherent in the furniture of the universe. We choose whole ecosystems in the boundary sense because managerial control of a geographically complete area is essential to protect larger contextual systems such as watersheds; environmentalists favor, also, protecting multilevel hierarchies—the complexity, integrity, and autonomy of all larger systems. But they do not insist on wholeness of all levels, except in the unusual experiment proposed at Yellowstone, an experiment that can be justified scientifically as providing a laboratory for studying the autonomous process of nature in evolutionary as well as ecological time.

Adoption of a pantheistic holism can now be seen as an intriguing option, a recognition that natural systems can be cathedrals as well as laboratories. But cathedrals represent human spiritual aspirations to recognize one's part in a larger whole, not an attribution of value to some supraorganismic whole that imposes value on human creatures. Contextualism is understanding human cultures in their larger, natural context; it does not require, as Chase suggests, management according to supraorganismic, human-independent, "metaphysical" values. Because it is that strong, metaphysical interpretation that leads to the charge of ecological fascism, which has been directed against certain "holists,"[54] environmentalists should choose metaphysical minimalism, and favor scientific contextualism rather than supraorganismic holism.

Integrated Land Management

At the end of Chapter 2 we half-jokingly suggested that Muir and Pinchot might have resolved their differences by partitioning their domains. Muir would be Minister of Aesthetics and would oversee and protect magnificent landscapes, while Pinchot, as Minister of Wise Use, would encourage conservation practices on lands already under intense economic use. Indeed, the idea behind Park Service and its careful bureaucratic insulation from the more commodity-oriented Forest Service can be understood as an attempt to forge just such a compromise between rapid economic growth and landscape protection.

But the partitioning did not work; indeed, it could not have worked. The isolationist approach of early preservationists did not take proper account of the effects of visitors and impacts of development outside park boundaries. More importantly, the partitioning was bound to fail because, in the rapid expansion of economic uses of the land, conflict over uses of particular lands inevitably occurred and these intensified as development continued. Unfortunately for the compromise by partition, lands in their natural state do not come with labels "to be preserved" or "to be used wisely."

Leopold is so important in the history of environmental thought and, especially, land use policy because, rather than *deciding between* the criteria of Pinchot and Muir, he *integrated*, at least theoretically, the diverging concerns of the two early approaches to environmental protection. Leopold insightfully recognized that, if humans are to be truly integrated into their environment, then their day-to-day activities—economic and otherwise—must be seen *as a part of the larger landscape*. Preservation cannot be accomplished by isolation, by locking up areas and eliminating human impacts, nor can conservation be accomplished by atomistic management, by maximizing productivity with no concern for the larger context of those activities. The diverse values of the North American people can be protected only in a patchy landscape, which mixes intense-use, limited-use, and natural areas that are integrally related in a larger environment.

The Yellowstone area, vast and unspoiled, but facing intense pressures toward economic development, may provide a laboratory in which the emerging consensus can be given a concrete form. The drought-stricken summer of 1988 brought devastating wildfires to the Yellowstone region; 988,975 acres within the park—44.5 percent of the park's total acreage—were burned. Intense political pressure eventually forced the Park Service to fight the fires more actively, but by that time the fires were too intense and widespread to yield to human efforts. Americans watched, on every evening news for months, their most magnificent park going up in flames. They also watched highly polarized debates, on "Nightline" and other popular programs, pitting natural-burn theorists against the advocates of the more traditional attitude, associated with Smokey the Bear, that the only good forest fire is an extinguished forest fire. Now that the smoke has cleared, it is possible to see some truth in the emotional outpourings from both sides. The Yellowstone fires underscore the importance of *historical context* in framing current management decisions. The fires of 1988 were not "natural." Indian tribes burned the area extensively until they were removed from the park in the 1870s.[55] Park managers suppressed all fires, as possible, for almost one hundred years. The result was huge expanses of over-aged lodgepole pine forests, full of dead wood and other hot-burning fuels.[56] The fires of 1988 were so large and so intense because of prior human alterations of the system. Those who wished to defend the let-it-burn policy of park managers as simply letting "nature" take its course, therefore, must ignore thirty thousand years of tribal history and one hundred years of Park Service management policy. On the other side, however, advocates of a return to the Smokey-the-Bear-stamp-out-all-fires approach must live with the embarassment that one hundred years of their policy created conditions ripe for uncontrollable fires. It is not clear that, even if the Park Service had responded in force against the earliest fires, the outcome would have been much different, given artificial conditions.

What this emotion-charged example proves is that the ideal of natural regulation is an empty ideal, unless it is understood within a larger, historical context.[57] Science is important, just as natural values are important, but science cannot turn back the hands of time. Natural regulation can be "natural" in the sense that it promotes wildness and autonomy, by degrees, in protected natural systems. But that ideal is a cultural creation built upon, and in, human culture. If the ideal of natural regulation is divorced from who we are, where we come from, and where we are going, it becomes a dead ideal, not capable of functioning at the dynamic interface of natural and cultural systems. This is the meaning of Leopold's conclusion that the land ethic must be viewed as a cultural creation.

Management of parks to maximize wildness exemplifies the value we place on natural, autonomous systems. But natural regulation is not a *rule*, it is an *ideal*, which is to be applied *where possible*. Ecological experiments are an essential park of understanding natural systems. Management "experiments," however, are better understood as expressions of the aesthetic creativity of a culture. In democratic societies, this means building the political will to develop a conception of the good life that is sensitive to its natural and historical context. This is what Leopold meant when he described the task of developing parks for recreation as "a job not of building roads into lovely country, but of building receptivity into the still unlovely human mind."[58]

From Leopold's perspective, environmentalists could learn much by listening to Chase's recipe for building a consensus for Yellowstone management goals *locally*, in the Yellowstone area. Chase suggests, for example, that park managers replace a dogmatic application of natural regulation with an educational outreach program. "Ordinary people," Chase suggests, "not just environmentalists and bureaucrats, must become part of the process."[59]

A dialogue based on mutual respect, with local residents expressing their views about the natural and cultural heritage of the region—as well as their economic aspirations—and environmentalists expressing concerns for the larger contextual values—the whole—is the only means to protect the cultural and ecological character of the area. Park Service personnel should act as facilitators of this dialogue; they should also provide scientific and managerial information to enlighten the discussion. Local environmental groups should accept responsibility to emphasize the local advantages of recognizing and respecting ecological and cultural limits embodied in the local land communities. If holism, with its implication of top-down values, is replaced by contextualism—understanding human activities as part of a larger, hierarchically organized whole—local values will be integrated into a larger, optimally healthy, environing system.

Complexity, diversity, and local heterogeneity cannot be enforced from a centralized vantage point. Ecology implies as much: Diversity is created as species adapt to local niches, microhabitats. The goal of land use planning should be to create means by which local communities can creatively express their particular natures. Local values can be protected, however, only as part of a larger context, a vast continentwide mosaic composed of complex, overlapping management cells and their environing systems.

Part Three

Environmental Philosophy

10

Diverging Worldviews, Converging Policies

The Emerging Consensus

This book began with an anecdote, my encounter with an eight-year-old with hundreds of living sand dollars. While I knew what I wanted the little girl to do—I wanted her to put most of the living sand dollars back in the lagoon—I felt in a quandary when I tried to explain *why* she should do so. I had no objection if the little girl took a couple home, to watch them in her aquarium or even to dissect them to learn their structure. But the family's actions showed no respect for life or living systems. I wanted to make a moral point not expressible in the language of economics. I hesitated to introduce, however, without serious qualifications, the moral language of rights. Rights have an individualistic ring about them; if sand dollars have rights, then surely the family should put them *all* back. One language said too little, the other said too much.

This original intuition, that the environmentalists' dilemma is mainly a dilemma of values and explanations, more than preferred actions, has been borne out by the considerations of the second part of this book. An examination of major areas of environmental policy has reinforced the hypothesis that a consensus on the broad outlines of an intelligent policy is emerging among environmentalists, even though there remain significant value differences that affect the explanations and justifications they offer for basically equivalent policies.

Environmentalists of different stripes, as far back as the days of Pinchot and Muir, have often set aside their differences to work for common goals,

187

But those traditional cooperations were, it seemed, almost accidental collaborations originating in temporary political expediency. My hypothesis about the current environmental scene asserts a more than accidental growth in cooperation: In spite of occasional rancorous disputes, the original factions of environmentalism are being forced together, regardless of their value commitments.

For example, a growing sense of urgency led soil conservationists and preservationist groups to work together to pass the 1985 Farm Bill, even though they suffered some ill feelings along the way. Similarly, the National Wildlife Federation, a collection of sportsmen's organizations, and Defenders of Wildlife advocate similar wetlands protection policies. While the value they place on wildlife is very different, the policy of protecting wildlife habitat represents a common-denominator objective, and the National Wildlife Federation is an effective lobbier for legislation to protect nongame endangered species as well as game species and their habitats. Given our present scientific knowledge about wildlife populations, hunters and animal protectionists alike conclude that we must aggressively protect the remaining habitats for migrating waterfowl. Both groups would also agree on the importance of careful management to protect the reserves from the effects of human activity. This consensus signals the end of both the atomistic style of single-species game management characteristic of early conservationists and of isolationist preservationism, which in its extreme form repudiated management altogether.

Several gradual changes undermined the two extremes of management style. Not the least of these causes was the progressive development of the nation, which increased the likelihood of spillover effects of one activity on another. Another cause was the rapidly increasing demand for outdoor recreation that began in the forties and fifties and has developed steadily since. In general, as population expanded and diversified, more demands were put on more lands, and decisions to use land for productive purposes led to more and more direct conflicts. Leopold's land ethic, which recognizes that the land community is a larger system in which human activities must be integrated, has led environmentalists beyond both atomism and isolationism.

What this means, in more concrete terms, is that all environmentalists, regardless of their allegiance to diverging traditions, must seek to manage the entire mosaic that is the American landscape. If we are to maintain the productivity of American agriculture *and* protect biological diversity, if we are to maintain adequate water supplies for homes and industry *and* preserve some wild and scenic rivers, if we are to provide sufficient opportunities for outdoor activities *and* preserve the pristine nature of wilderness areas, we must make large-scale land use decisions with an eye to their larger context. A landscape that can accommodate all of the varied aspirations of Americans will have to be a patchy landscape, in which urban elements, productive elements, and pristine elements are arranged intelligently. Each of the patches must be managed according to the methods appropriate to goals that define its use, but those methods must also be designed to enhance, or at least not destroy, the values sought elsewhere in the mosaic. Further, the principles of this holistic management must be aesthetic as well as economic, and historically informed as well as forward-looking. But they must be applied to the entire context of human activities, not to specific activities viewed either atomistically or in isolation from other activities.

The forced abandonment of the two extreme styles of management associated with the old split between conservationists and preservationists provides a useful first stab at characterizing the emerging consensus. Surveying the results of the last four chapters, we see a pattern of emerging policies that unite environmentalists in opposition to the policies usually favored by production-oriented developers. In each area—resource use, pollution control, protection of biological diversity, and land use policy—environmentalists advocate limits that guide the search for acceptable policy options, insisting against the simple economic Aggregators that there are constraints governing human exploitative activities. These constraints are usually stated in terms of "sustainability," but Leopold and modern environmentalists have gone beyond demand-oriented conceptions of sustainability to recognize limits inherent in the complexity and organizational integrity of larger ecological systems. Similarly, in pollution policy, environmentalists have recognized constraints on activities that pollute the environment based on rights of other individuals to a healthy environment.

The common denominator of these obligations of resource users to limit their activities in these diverse cases cannot be understood as a commitment to any particular moral principle such as the moral equality of all species or of interpersonal equity. The common element is structural: in each case, individually motivated behaviors, which can be understood as activities of economic man, are constrained because of the impacts those behaviors impose on their larger context. Environmentalists emphasize total diversity and biological complexity because the complex processes that constitute biological systems *are* the larger context of all life, human and nonhuman. Rapid alteration of those larger systems will cause serious disruption of both human and nonhuman activities. Land must therefore be used according to patterns that protect the complex processes of nature, so as to avoid destabilizing changes, changes in environing systems that are too rapid to allow human activities and nonhuman processes to respond and adapt.

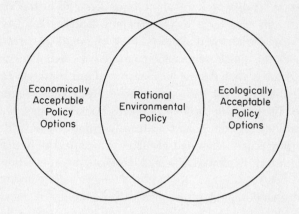

Figure 10.1. The politically desirable solution, according to contextualism, is chosen from the intersection of ecologically and economically acceptable policy options.

This consensus represents, in one sense, a victory of Moralists over Aggregators. The essence of the simple aggregationist approach was to reduce all questions in environmental policy to calculations regarding economic efficiency, to judge all questions on a single scale. Environmentalists have rejected simple aggregationism and insist on moral constraints ranging from individual rights to clean air to imperatives to protect the integrity of ecological systems. Pinchot, however, has left his mark as well. In the spirit of the great compromiser, environmentalists play pragmatic politics—they seek their policies from among the politically viable options. In political practice, both moral and economic imperatives exercise a veto power (see Figure 10.1).

In the context of political debate, individual rights, moral obligations to protect species, and scientifically articulated thresholds or constraints inherent in fragile ecological systems must all be factored into a process that sets goals, objectives, and standards for environmental programs. Economics are, of course, not irrelevant. Once goals and standards are set politically, it makes sense to use economic analysis to rank those alternative actions and policies regarding their efficiency in achieving the politically determined goal. Environmentalists know that economic interests will block environmental legislation if it is too costly. It behooves environmentalists, therefore, to propose the least costly option that will fulfill their goals.

Once the crucial distinction between setting goals and choosing means to achieve those goals is recognized, contextualism need not emphasize prohibitions and regulations in order to encourage actions that respect constraints. With an adequate conception of health for the overall ecological context in place, efforts can be shifted from prohibition and regulation to the creation of incentives that encourage individual actors to choose less polluting activities or to choose land uses that will help, rather than harm, efforts to protect biological diversity.

So, for all its emphasis on constraints, the emerging consensus among environmentalists need not operate, on a day-to-day basis, by constraining individual activities. To return to our example of the farmer who clears his woodlot to plant wheat, the emerging consensus need not, when trends in farmers' behavior begin to press against thresholds inherent in the land community, regulate the farmer. It may, instead, choose to institute incentives that will encourage some farmers, either the one in question or others, to let a wheat field go fallow. The beauty of contextualism is that, once ecologically informed constraints are formulated, the society can undertake positive steps to encourage individuals to act in ways that counter dangerous trends. By combining a positive, biological conception of a healthy ecosystem with a program of incentives, the emerging consensus offers an alternative to simple reductionistic economics on the one side and onerous restrictions on the other. Following David Brower, this alternative can be called "restorationism." A positive definition of ecosystem health, one that incorporates human activities as long as they do not threaten thresholds inherent in ecological systems, opens the possibility of a truly positive ideal of humans living, creatively and freely, but harmoniously, within a larger, ecological context.

According to the contextualist synthesis, resource management, in the tradition of Pinchot's wise-use conservationism, is a respected activity, but one that must be understood as limited by the broader, contextual constraints of environmental management. These constraints express Muir's insight that all elements of nature must be seen as ultimately interrelated. As we shall see in the next section, the utilitarian value system so useful to Pinchot and the more holistic ideas of Muir are still strongly represented among environmentalists. What has changed, as the movement has matured, is that these metaphysical and moral worldviews have found distinctive habitats. Wise-use conservation guides resource use policy and largely governs land use decisions on lands already devoted to intense human use. Contextualist thinking, which insists on seeing resource use activities as a part of nature's larger whole, concentrates on problems caused in larger systems in which resource-producing units, such as field or forest, are embedded.

What once appeared as a war between two factions with opposed worldviews now appears as two protective strategies that are applicable in differing situations. As long as the larger, environmental context remains healthy, resource use strategies will be guided by maximization criteria. When exploitative activities threaten to exceed ecological parameters—the fragilities inherent in particular situations—*environmental* managers urge the enforcement of constraints that will limit the resource-producing activities that threaten their environing systems. Because of scientific uncertainty and because the second-order criteria for deciding which lands should be protected and which should be used wisely are imprecise, environmentalists often disagree about how particular parcels should be used.

When environmentalists differ regarding policies and strategies today, they are usually disagreeing regarding the appropriateness of a given strategy to a particular situation, as, for example, when they disagree about which areas and how much National Forest land should be devoted to protection as opposed to production. These disagreements are often based on very different assessments of the magnitude of certain spillover effects of exploitation—such as stream degradation and loss of diversity—on their larger context and on the values humans derive from that context. Needless to say, self-interest also plays a role in advocacy of land-use decisions, as evidenced by the NIMBY (not-in-my-backyard) syndrome, in which local groups gain their power from quality-of-life arguments, but arguments that emphasize quality of life in very small contexts, such as a suburban municipality.

As these local disagreements continue, a further reach of consensus is emerging: even the American environment, taken as a huge unit, is not self-sufficient. As I interviewed leaders of the environmental movement, one point of consensus stood out, mentioned by almost all interviewees—the American environmental movement must, and will, become more international in the scope of its programs. Environmental groups that have been focused mainly on problems in the United States have instituted international programs, especially in the areas of biological diversity, but increasingly in other areas as well. Partly this change is a matter of the changing nature of environmental problems: Acid rain, the greenhouse effect, pollution of the oceans, and depletion of the marine fisheries do not respect national boundaries. But the urge toward internationalism in environmental affairs is also another expression of the emerging consensus—contextual management is management as part of a system; the system in which the dynamic American mosaic will evolve is the entire world.

Scientific Environmentalism

A broadly contextualist approach to environmental policy characterizes the emerging consensus among environmentalists, but the policy consensus is not accidental—it is forged on the basis of a shared view of science. This agreement is evidenced in tentative agreement on some basic scientific "axioms" to guide environmental management and, more profoundly, in a shared acceptance of science as a value-laden enterprise. The underlying scientific consensus can be summarized by listing and drawing out the consequences of five axioms and an associated definition of ecosystem health.

1. *The Axiom of Dynamism.* Nature is more profoundly a set of processes than a collection of objects; all is in flux.

2. *The Axiom of Relatedness.* All processes are related to all other processes.

3. *The Axiom of Systematicity.* Processes are not related equally, but unfold in systems within systems, which differ mainly regarding the temporal and spatial scale on which they are organized.

4. *The Axiom of Creativity.* The autonomous processes of nature are creative and represent the basis for all biologically based productivity.

5. *The Axiom of Differential Fragility.* Ecological systems, which form the context of all human activities, vary in the extent to which they can absorb and equilibrate human-caused disruptions in their autonomous processes.

These five axioms are basic elements of a worldview in the sense that they shape and give context to environmental science. While the most explicitly evaluative aspects of environmentalists' worldviews have remained divergent, these broadly scientific axioms function as a shared scientific component of an emerging worldview. These axioms, I submit, provide the scientific portion of a new and emerging worldview and shape environmentalists' conception of the problems of environmental management.

In practice, these axioms are applied in conjunction with a highly normative definition of ecosystem health: An ecological system is healthy only when its creative processes, represented by the free flow of energy and active competition to utilize it, remain intact. Unhealthy ecological systems will be characterized by a tendency to undergo rapid change, change such as the rapid disintegration of complexity and integrity that Leopold observed as grazing caused deterioration of fragile vegetative communities in the arid Southwest. This shared understanding of the problems of environmental management encourages an analogy between medicine and environmental management. Both are empirical arts; a physician would never administer a heart drug without concern for its effects on the kidneys or on the overall health of the patient. Similarly, resource and environmental managers should practice preventative medicine, moderating dangerous trends and avoiding disruptive "illnesses" that they do not know how to cure.

Environmental managers differ from physicians as well. Since the study of whole ecosystem management is in its infancy compared with medicine, the dangers of iatrogeny (illnesses caused by medical treatment) are great.

The treatments prescribed by resource and environmental managers must be carefully considered for their effects on the whole.

The creative processes of nature, as perceived within this worldview, are understood essentially as energy flows up the biotic pyramid. The diversity and complexity of nature result from varied adaptations of systems that exploit that energy. The sun supplies energy and local variation provides the context for ecological and evolutionary processes. Freedom, in an ecologically organized world, represents the creative ability to react to environmental variations, to adapt to an opportunity in the pattern of flows in the energy pyramid. Each species carves a niche in an ecological system either by coevolving with other species or by invading or colonizing a new habitat. The adaptive "choices" species make, the apparent freedom of species to react differentially to varied ecological opportunities and challenges, represents the "freedom" inherent in natural systems. This metaphysical concept of freedom provides the basis for the more inclusive concepts of "integrity" and "autonomy." A system maintains its integrity and autonomy when its future states are a result of ecological processes, interactions among species in ecological time. Human activities that use powerful technologies to alter systems rapidly tend to disrupt these normally slow-changing interactions.

Central to the emerging ecological consensus is the Axiom of Dynamism and its complex interplay with the other four axioms. Because, in the environmentalists' worldview, all is in flux and, according to the Axiom of Differential Fragility, treatment of the environment must be tailored to local conditions, it follows that values (whatever their ultimate formulation) will be given operational meaning in changing local contexts. Values cannot therefore be conceived as static and unchanging—they are given meaning at the sharp cutting edge of the search for a culturally, politically, and environmentally viable conception of the good life.

Because human values are a part of the ever-changing dynamic system, values cannot be conceived as static and unchanging preferences, as simple Aggregators interpret them. The static conception of human values as unquestioned preferences cannot express Muir's insight that observation and scientific understanding can have an ecstatic aspect, causing a shift in worldview toward less consumptive and more contemplative pursuits, for example. Environmentalists conceive environmental management to have public education as one of its central goals. Environmental managers do not have a ready-made, a priori criterion to guide them in the search for a healthy environment. They must engage the public in a dialogue, a political process, of defining such terms as "ecosystem health," "ecological restoration," and "ecosystem integrity." As scientific understanding of the larger, ecological context improves throughout the population, environmentalists believe, citizens will begin to see the world in a new, more holistic light, and the result will be a change in values. This is the transformational element in the search for an adequate ethic for dealing with human impacts on the natural world.

The Axiom of Systematicity, which interprets nature as systems embedded within systems, and as changing according to multiple frames of time, rules out any simple form of aggregationism as the basis for determining environmental policy. Some events will be given meaning, on the contextualist worldview, in smaller systems, others will be interpreted as events in larger systems, and still others will be given meaning in more than one system. One cannot meaningfully calculate the value of events that have different meanings in these different systems within a single system of value such as that of economics. Ecologists, by providing information about what is possible in management and what is not, participate as favored advisors in the decision process. It is from their point of view that holistic constraints, based on projections of change in contextual systems, are formulated. Here, ecologists and conservation biologists adopt tasks analogous to those of physiologists and physicians when they enter public debate about the meaning of, and criteria for, human health.

As citizens begin to see their behavior in an ecological context, there will be a shift toward the ecological worldview, and management problems will be seen as scientific problems in the sense that science, especially ecology, determines the context in which resource use questions will be addressed. Goals will be formulated in ecological language; this means that some objectives will be justified simply as protecting functioning systems and their organization. In this sense, an environmental ethic is "holistic." Public discussion of the goals of environmental management will concern the search for the good life in a good environment, and a good environment presupposes stable, slow-changing environing systems.

The language, environmentalists agree, in which this discussion should be articulated is the language of ecology supplemented with ethics, not the language of economics. Environmentalists are united by their belief that ecological science is capable of transforming worldviews and ultimate values; we saw in Chapter 2 that both Pinchot and Muir thought of themselves as scientists, but that they practiced science in two distinct traditions. Pinchot saw scientific resource management as pursuing socially determined values; he avoided, to the extent possible, value judgments in his work. Muir, however, doubted that science can ever be a value-free enterprise, ridiculed what he saw as the futile and self-defeating attempts to observe the world "objectively," and recognized that the basic assumptions one makes about the order and structure of nature are value-laden. Because the sensibilities of environmentalists evolved within the naturalist tradition, in which the works of scientists, artists, and poets were not sharply separated, naturalists are comfortable with a vocabulary sprinkled with openly evaluative terms, including "healthy ecosystem," "ecosystem integrity," and so forth.

Pinchot, on the other hand, embraced positivistic science on its upswing. For the first half of the twentieth century, objectivity and value-neutrality were the watchwords in the scientific community. Muir's ecstatic science was eclipsed. In the aftermath of the loss of the Hetch Hetchy controversy, Muir's death, and the inexorable growth of government bureaucracies to encourage wise use of resources, the naturalist tradition was unceremoniously dismissed as unscientific.

Only recently has positivism lost its grip on professional philosophy, but it remains strong in many of the special sciences. Popular among philosophers in the 1930s, 1940s, and 1950s, the positivist view is now seen as seriously oversimplified. Philosophers still follow positivists in recognizing the importance of falsifiability in science, but they now emphasize that "falsification" always takes place against a backdrop of theoretical assumptions. Theoretical assumptions are often affected by the values that we hold. It follows that scientists work, even when testing hypotheses, in a world permeated by values. Environmentalists have, in this respect, chosen Muir's worldview over Pinchot's. In the words of Leopold, they have adopted ecology as "the new fusion point of the sciences."

It now becomes clear why it is so difficult to separate scientific, descriptive elements of a worldview from its evaluative components. The very conception of the world as a set of systematic processes that evolved to manage and allocate energy implies that, within the ecological system so described, there is value in maintaining those organizations. Neither economic productivity nor aesthetic and intergenerational moral values can be protected without protecting the complex, organized system that provides the ecological context on which *all* values depend. To assume that all values are economic values is to ignore this implicit, background value in the ecological processes that support economic, and all other, activity.

A contextualist science of management is value-laden in yet another respect. The choice of a scientific management unit—boundaries for a system to be given a "management plan"—is inseparable from moral, aesthetic, and cultural choices regarding goals of management. One cannot, for example, decide on the appropriate management unit for the Chesapeake Bay without considering how important protecting bay water quality is in the management plan; similarly, the choice to manage the GYE for wildness indicates a management unit sufficiently large to contain major migrations of wildlife populations. Choices of the scale on which to conceive management problems are inseparable from the adoption of appropriate management goals.

Environmentalists who endorse a variety of values will nevertheless tend to see environmental problems in a similar way and will pursue common-denominator goals, even though they might explain and justify those goals in quite different evaluative language, because the larger context of environmental management—the context in which all values are pursued—will be understood as the larger ecological context. Environmentalists have been able to fashion a working consensus for addressing environmental problems on an ecological basis precisely because they understand the world as the context of multiple values. This understanding unites them behind goals essential to protect a wide variety of values, however expressed, but the impetus toward the consensus is scientific. Environmentalists are being driven together by their commitment to ecological contextualism, which implies that all human values depend upon a healthy context.

The Worldviews of Environmentalists Revisited

This growing scientific consensus has not been accompanied by a corresponding narrowing of the value positions that environmentalists employ. We can separate out at least seven worldviews or fragments of worldviews, all of which are sufficient to trigger, depending on the situation, a constraint on exploitation and development. While each can stand independently against the reductionist approach, environmentalists often appeal to a shifting mixture of all seven worldviews. These are

1. *Judaeo-Christian stewardship*. While some environmentalists are critical of the human-centered emphasis of Genesis I, many Christians and Jews accept a form of benign dominion. This view emphasizes the admonition of Genesis II, that humans were put in the garden to "dress and keep it."[1] Environmentally aware Christians and Jews can, from this biblical admonition, justifiably urge that each generation pass the ecological context on to the next, intact in its creative complexity. This tradition of Christian stewardship recognizes obligations to the Creator not to destroy his handiwork. It recognizes, in other words, theologically based obligations not to destroy ecological systems or the species that compose them.

2. *Deep ecology and related value systems*. These systems are characterized by attributions of *rights* to nonhuman nature or by the related belief that nonhuman elements of nature have *intrinsic value* and that this intrinsic value places obligations and limits on the extent of human prerogatives to use and alter nature.

3. *Transformationalist/transcendentalism*. Within this framework, wild

nature has spiritual value because experience of nature can transform human perception and value. Nature becomes, in this worldview, the cure for human alienation caused by modern society, a sacred space that humans enter to reconsider and reform their worldviews and value systems. Although it is possible to link a transformationalist approach with a belief that nature has rights—the transformation may be a recognition that nature is intrinsically valuable—this linkage is not necessary. If shifts in worldview are valued as improving human values and satisfactions—stamping out "quiet desperation" in the immortal words of Thoreau—then this moralistic position is anthropocentric and spiritually instrumentalist.[2]

4. *Constrained economics.* Many environmentalists continue to perceive the problem of resource use as essentially a problem in human economics, but for a variety of reasons they reject Aggregationism. A good example of this approach, encountered in Chapter 8, is the Safe Minimum Standard (SMS) approach to protecting species and natural systems. On this approach, actions to avoid irreversible damage to the environment are assumed to have great, though not quantifiable benefits; these actions should be taken as long as the costs of doing so are not unacceptable. These constraints on economic activity might be justified by appeal to other value systems, such as (1) and (2), but need not be. They might, instead, be based on common sense and a relatively low threshold for risk-taking.

5. *Scientific naturalism.* Scientific naturalists are characterized by their broadly Darwinian view of life's processes and by an associated emphasis on dynamism and contextualism. Naturalists tend to see problems, including resource use problems, scientifically, but they do not accept the economists' suggestion that policy be determined by economic calculation because they believe that a broadly ecological/evolutionary worldview implies contextual limits, limits on population growth and violence to the land. These constraints follow, they infer, from the ecological-evolutionary conception of a species and its niche.

6. *Ecofeminism.* A significant group of feminist scholars argue that the domination of nature is symbolic of gender domination more generally.[3] They oppose the positivist worldview as an ideology of domination and see simply aggregative economics as disguised ideology. They argue that rejection of the ideology of domination will encourage solutions to both environmental and social problems.

7. *Pluralism/pragmatism.* Activists who are not philosophically inclined, and philosophers such as pragmatists who doubt the efficacy of general, overarching moral principles and theories, gravitate toward a pluralistic approach to environmental values. On this approach, practical problem-solving and ethical ideas and principles are enlisted, not so much as a priori principles that enforce consensus, but as useful means to recognize similar

features of varied cases—as useful tools, in other words, to aid in the development of a solution to moral quandaries.

Can, or should, these various worldviews of environmentalism be unified? Moral monists, as they are described by Christopher Stone, share two tenets: (1) that the goal of moral reasoning is "to produce, and to defend against all rivals, a single coherent and complete set of principles capable of governing all moral quandaries"; and (2) that this is a determinate goal in the sense that the favored framework "is to yield for each quandary one right answer."[4] I have argued that, empirically speaking, environmentalists have been pluralists, described by Stone as exhibiting a willingness "to develop a conception of the moral realm as consisting in several different schemata, side by side."[5]

Many commentators imply that environmentalists must, in some ultimate sense, be monists. When Roderick Nash, for example, implies that the history of environmental ethics is the history of the single idea that nonhuman elements of nature have rights, he implicitly assumes moral monism, implying that all goals environmentalists pursue must be supportable on a biocentric principle.[6] This view is no less reductionistic than the monism of economic Aggregators. The moral monist's method is to sort through moral maxims and hold them up against a single measure. If they can be derived from the central principle, they are subsumed under it as corollaries. If not, they are rejected. By this method, the monist would reduce the seven worldviews and fragments to one—the "correct" environmentalists' worldview. Monists, in other words, place a high value on connectedness and pursue the goal of stating a single, rock-bottom principle to guide all moral action.

The strategic question of whether it would be politically useful for environmentalists to have a single worldview must of course be kept separate from the epistemological question of whether there is a correct worldview. The strategic question will be addressed in the second section of Chapter 12. Here, we may ask whether environmentalists are dishonest, or otherwise in error, when they use a variety of worldviews to justify their policies. Nash clearly implies that environmentalists' pluralist tendencies are, at the least, devious. Nash describes Muir as fully committed to the rights of all nature by 1867,[7] but "Muir knew very well that to go before Congress and the public arguing for national parks as places where snakes, redwood trees, beavers, and rocks could exercise their natural rights to life and liberty would be to invite instant ridicule and weaken the cause he wished to advance." Nash concludes that Muir therefore "camouflaged his radical egalitarianism in more acceptable rhetoric centered on the benefits of nature for people."[8]

Nash's attitude is further revealed when he compares Joseph Wood

Krutch's philosophy of environmentalism with Leopold's: "Like Leopold, Krutch left some ambiguity as to whether his extended ethic was *pure* [my emphasis], in the sense of respecting the rights of other parts of existence, or instrumentally to the successful continuation of human existence. Pragmatically, he knew, as did Leopold, that the latter had a better chance to win public favor in the 1950s." This concern for the "purity" of biocentrism is revealing. Even while recognizing that "the main thrust of his [Krutch's] philosophy, as of Leopold's, centered on the idea that an ethical attitude toward nature was 'better in the long run for [humans] also,' " Nash insists that any public reference to human interests in preservation is cheating; they are based on political, rather than moral, goals and are therefore not "pure."[9]

In fact, we have seen, the environmental movement has been pluralistic in its value commitments, given to political compromise, not ideological exclusivity; neither simple aggregationism nor monistic biocentrism could, by itself, express both the wise-use concerns of Pinchot's followers and the systematic constraints characteristic of contextual, ecological thinking. It is moral monism, of course, that gives teeth to the environmentalists' dilemma by encouraging a perception of human utilitarian and nonanthropocentric values as exclusive. The pluralist, on the other hand, can search for value in nature that is reducible neither to the dollars of economists nor to the rights of wild species. The moral pluralist can look for common ground from which to construct a new, philosophically, culturally, and politically viable worldview that sees humans as integrated into larger systems and that values objects as parts of their human, cultural, biotic, and abiotic contexts.

But there are also difficulties in accepting pluralism.[10] To simply say that different principles of value apply in different contexts introduces moral chaos unless something is said about which particular principles apply when. Most of the important disagreements among environmentalists—the Hetch Hetchy controversy, for example—have revolved around which criterion should govern particular policies and land uses. Pluralism can provide guidance in environmental policy only if it includes second-order principles that help to determine which of its diverse first-order moral criteria apply in given situations.

A pluralistic system with such second-order principles could be called an *integrated worldview*. It would be integrated in the sense that each of the principles would be given an appropriate domain of application, according to second-order rules based on a determination of the context of the managerial problem faced. It would also be integrated in a deeper sense, provided those rules are interpreted systematically. A truly integrated system of thought, an adequate environmental worldview, would state rules of application according to the systematic context of the management problem faced. The criterion, according to a contextual approach such as Leopold's land ethic, should be based on the temporal and spatial scale appropriate to the problem at hand.

Environmentalism and Naturalism

It is time to think seriously, if briefly, about ethical theory. Having argued that environmentalists agree on what to do *because* they agree on some basic scientific, especially ecological, axioms, it appears that we are flirting with the dread "naturalistic fallacy." Philosophers, having noted that factual statements—which describe the way the world *is*—differ in their logic from evaluative statements—which prescribe the way the world *ought* to be— have generally subscribed to David Hume's "Gulf Doctrine." According to the Gulf Doctrine, as stated by Hume, any expression containing an "ought," "expresses some new relation or affirmation," and cannot be derived from any number of factual propositions.[11]

If environmentalists are attributed the view that we should follow certain policies, justifying these policies by reference to axioms and theorems from the scientific study of ecology, they would appear to have traversed Hume's impassable Gulf. Must environmentalists, then, be ethical naturalists, and insist that questions of value will ultimately be determined by scientific progress?

As an aid in answering this abstract question in the logic of ethical reasoning, let's look again at the case of wetlands protection in the Chesapeake Bay region. I have chosen this region to keep matters as concrete as possible, but our considerations will be sufficiently general to apply to wetlands in New Jersey or South Carolina, or for that matter to wetlands in any of the great flyways for migratory waterfowl such as the Mississippi flyway or the Pacific flyway. Consider the incredible diversity of the groups that work for protecting wetlands habitats in these corridors; they include hunting and fishing organizations such as the National Wildlife Federation and Ducks Unlimited, birdwatchers such as local Audubon societies, local and statewide nongame wildlife groups, and organizations with roots in the humane tradition such as Defenders of Wildlife—a rather remarkable coalition.

Nor does the cooperation of these groups represent an accidental concurrence of opinion, a passing coalition of groups with usually divergent interests. These groups have worked for the same goals for decades and, while they espouse radically different values and use rhetoric offensive to each other, they can be expected to continue to support into the indefinite policy future the expansion of wetlands protection, more stringent limits on activities that destroy wetlands, and restoration of degraded wetlands. According to the general picture we have painted of the emerging consensus among environmentalists, a generally ecological argument has driven these diverse groups into the same policy camp; whether one likes wildfowl to shoot at or to look at, whether one touts the rights of birds, or whether one sees migratory waterfowl as an important part of a "whole" ecological context, our expanding understanding of ecological systems enforces on all of these groups the common-denominator objective of protecting and restoring wetlands habitats on the flyway corridors.

In our earlier methodological interlude (the second section of Chapter 5), we noted how ethical discussions, when healthy and open, tend to go around in circles. Philosophical discussions, in "pure ethics," go around a circle that starts with theory and tests that theory by cooking up constrained examples such as the fortress example, testing ethical theories at the limits of their application. Practical ethics, a messier business, goes around the circle by successively formulating theories and attempting to apply them to real-world cases. The reason for this circular pattern is that, while we often have an intuition about what is "right" in particular cases, we often do not know how to justify that intuition until we adopt a theoretical language for articulating our reasons; as we articulate our reasons within a theory, however, we often find ourselves accepting a general theory that would force us to give up other, deeply felt intuitions about other particular cases. In this way, the task of moral enlightenment resembles the task of Sisyphus, who was cursed to push a boulder up a mountain, only to have it slip back each time he reached the top. We avoid the depressing conclusion by hoping that by inching the boulder around and around the mountain, carving out footholds in theory and showing how they increase the stability of the whole enterprise in the process, we can spiral up the mountains even as we improve our understanding of both theory and individual cases.

Environmentalists, in the case at hand, continue to articulate differing justifications for saving wetlands habitats, and seem unlikely in the near term to reach consensus regarding the evaluative aspects of a worldview, even though their shared scientific understanding leads them to work toward the same policies. Do environmentalists such as hunters and humane advocates, who accept differing, even contrary, ultimate values therefore end up determining *what to do*—which clearly involves a value component— by reference to scientific, factual statements? Is the proposed consensus based on indefensible and fallacious naturalistic reasoning?

I think not but the question is sufficiently complex to justify a careful answer. The emerging consensus is unquestionably naturalistic in the broad sense that science informs and constrains decisions about what to do in these cases, without dictating specific values and in the sense that a consensus in policy emerges within an inductive debate concerning scientifically determinable local conditions. Ecology and ecological reasoning set the context for ethical debates. Information about the crucial role of wetlands in absorbing nutrients and limiting algae growth, information showing the crucial role of submerged aquatic vegetation in supporting migrating waterfowl, and facts about the importance of wetlands for migratory patterns generally—all focus attention on the policy goal of wetlands protection. Ecology therefore directs environmental concern to the *systematic* level, focusing attention on protecting whole complexes of wetlands. National and local groups espousing a wide variety of values and worldviews are therefore focused on the importance of habitat protection, even while some are polishing their field glasses and others are cleaning their shotguns. Scientific understanding of the ecological context affecting all of their diverse values forces them to adopt a common-denominator objective, wetlands protection.

Recognizing, however, that ecology forces these diverse groups to support the same policies does not imply that values are logically derived from facts. Indeed, the emerging consensus, I have hypothesized, may continue indefinitely even as values placed on waterfowl remain divergent. The consensus does not fallaciously derive values from facts; it is rather that ecology sets the context in which advocates of different values debate regarding policies. Whether one likes to shoot birds or look at birds, one is forced by our scientific understanding of how ecological systems sustain themselves through time and support wildlife—including temporary residents such as migratory birds—to look at a more systematic level of ecosystem organization in order to articulate a policy. And, on that level, one need not determine the specific value given to individual ducks or geese—a common-denominator objective of protecting habitat emerges because that habitat embodies the context in which multiple and even conflicting values are enjoyed.

An ecological and systematic understanding of environmental problems therefore supports a policy consensus without dictating individual values. Deeper scientific understanding, as yielded by ecology, of the context of environmental policy questions drives together participants in the debate. Successive trips around the circles of practical ethics show that, given improving understanding of the nature of environmental problems, one arrives at the same policy recommendations even while, on the theoretical points on the circle—the points at which participants are articulating ultimate values as a part of their worldview—considerable divergence remains. They reach policy consensus mainly because they agree about science, not because they agree about values.

Now, it might be argued that this situation is unstable, that the policy consensus will collapse unless environmentalists ultimately achieve a consensus in worldviews as well as policies. Indeed, some advocates of an "ecological worldview" would argue that a proper understanding of our place in the biotic system will ultimately determine the values we hold; that an understanding of ecology and evolutionary theory will lead, ultimately, to a rejection of hunting, to vegetarianism, and to a radical revolution in how we treat not just natural systems, but also individual members of wild species.[12] Those who believe ecology will have such a unifying effect on all aspects of our worldviews in the future seem to be closer to committing the naturalistic fallacy than the view sketched here in describing the general scientific basis of the emerging consensus. Whereas this sketch shows how individuals who espouse differing ultimate values will nevertheless converge on particular policies, scientific understanding demonstrates to them that, whatever their values, a common-denominator objective of habitat protection is essential to protecting their diverse values. The view that ecological understanding will eventually enforce on all discussants a univocal worldview seems to me not entirely implausible, but that strong version of naturalism is unnecessary to support the solid contextual consensus that is emerging among environmentalists.

11

Intertemporal Ethics

Looking Forward

"What good is a worldview, anyway?" we might well ask, if environmentalists are allowed to put them on and take them off like hats. This is serious business; after all—it's no fashion show—the future of the planet is at stake.

We have noted that environmentalists lack a fully developed worldview, a complete conceptual, theoretical, and evaluative framework for interpreting the world.

Environmentalists have generally, as David did in facing Goliath, gone into battle against the powerful forces of exploitation, which are well armed with a reductionistic worldview, with just slingshots and pebbles. But environmentalists have done remarkably well, given the apparently uneven distribution of intellectual armaments. The hit-and-run tactics of guerilla warfare have obvious benefits. Playing fast and loose with metaphysical and moral principles, environmentalists have gained considerable political clout by employing that value which seems particularly appropriate for a given issue, or by emphasizing a particular worldview that will be effective in reaching a coveted constituency.

But guerilla warfare has important costs as well. Environmentalists can appear to outsiders as disorganized and fractious, especially if one listens to their rhetoric, rather than observing their political actions. Further, the fragmentation of environmentalists' worldviews has real costs internal to the movement because it results in failures of communication and mistrust, even among individuals and groups that are pursuing identical or nearly identical policies. For example, while committees formed by the Group of Ten could reach a detailed consensus on policy in all areas of environmental concern, they were unable to present the document as endorsed by their respective organizations because some organizations wished not to be publicly associated with others because of differing attitudes toward hunting.

The most important cost of worldview fragmentation among environmentalists, however, exists not in the past or in the present, but in the future. Environmentalists have failed to articulate a positive vision for the future; they cannot explain in terms comprehensible to each other or to the public at large what is their positive dream. As is sometimes said, environmentalists are always "against something." This role as constant opposition stems, as we learned, from the roots of environmentalism: the movement, from the beginning, has reacted against the dominant forces of exploitation and development. In the beginning, this was no disadvantage. Just as guerilla warriors with quite different political values can unite to topple a corrupt and unpopular dictatorship, environmentalists, as long as they operate in an opposition role, can find unity in what they are against.

But, just as guerilla warriors often find it difficult to build a cohesive society and government once the established regime has been destroyed, reactivist environmentalism will fail to institute positive policies and programs when it has attained significant power. The single greatest weakness of the environmental movement today is its failure to project to the American population as a whole a coherent and plausible vision for the future, a picture of what it would be like fifty years, or a century, or five centuries in the future for humanity to live in harmony with the rest of the natural world. It is this failure, which is closely related to the failure to reach consensus on a shared worldview, that weakens environmentalism and leaves it unable to adopt a position of true leadership in public policy. What are the prospects for environmentalists to fill this leadership void, by developing and projecting into public policy discourse a unified vision, a worldview capable of capturing the imaginations and dreams of all Americans?

In searching for a more comprehensive vision for the future, let us return to the fruitful ideas of contextual management, which looks at individual actions and trends *from a local viewpoint*, but *within their larger context*. On this approach, actions and policies that are benign in many situations fall subject to moral constraints when trends in those activities threaten their larger context. This insight may provide the basis for a principle of integration by indicating conditions under which moral constraints take priority in given situations.

When some initiative in environmental management goes awry, as Leopold learned, it is often because that initiative was undertaken without proper attention to the larger context within which the cell isolated for management is embedded. Taking a clue from hierarchy theory, we can further hypothesize that management failures will usually exhibit scalar qualities of two general types, spatial or temporal. Let us begin the search for a more integrated, but pluralistic, worldview by noting that the particular moral values environmentalists marshal in a given context may depend on whether the failure in question forces a look at impacts across longer frames of time, or as these impacts ripple outwardly, affecting other species and their habitats. Too narrow approaches fail to consider all of the impacts of a given action on its larger physical context; shortsighted approaches fail because they did not foresee, at some point in the past, distant effects of actions assumed at that time to be environmentally harmless. Contextual management helps us to avoid atomistic approaches and to look at management problems always with one eye on the larger system as it changes through time.

This emphasis on scales suggests that we may need two quite different ethical scales—one for analyzing our actions as they unfold in larger and larger scales of time and another, as they ripple outward through more loosely connected systems. Analysis according to the intertemporal scale can be called "diachronic."[1] Another moral scale would apply as their effects ripple outward into other systems viewed "synchronically," at a fixed point of time. In the remainder of this chapter, accordingly, we will examine environmental problems diachronically, in larger frames of time. Chapter 11 will examine contexts of environmental problems synchronically, in expanding physical contexts.

Third-Generation Environmental Problems

Environmentalists have not faced a single, unchanging pattern of problems over the past century. As a reactive movement, they have been regularly forced to adapt their policies, and even their ideologies, to changing conditions. For example, Rachel Carson's *Silent Spring* is often seen as a herald of modern environmentalism, a shift away from concerns for wise use of resources and protection of particularly spectacular natural monuments toward the more pervasive and less focused problems of pollution, biological simplification, and regional deterioration due to thoughtless land development. Robert Mitchell has therefore distinguished second-generation environmental problems from the earlier first-generation problems.[2]

Recent headlines have underscored yet another shift in the challenges modern industrial societies pose for environmental protectors. Fears of ozone depletion, of the greenhouse effect, of acid rain, and of an abrupt and cataclysmic loss of biological diversity form an emerging group of problems that bear little relation to first-generation problems. But these more newly recognized environmental problems share a complex of features that distinguish them from second-generation problems as well. In each case a catastrophic effect, resulting from a large number of small but incremental decisions made in the present, would affect mainly future generations. These catastrophic effects would harm large numbers of people over geographically broad areas. While these effects may appear unlikely if the historical record is taken as a base, fear of them nevertheless seems rational because respectable scientific models predict serious consequences that would be irreversible.

These newer problems form a subset of the problems that have been classified by Talbot Page and Ezra Mishan as "zero-infinity dilemmas.[3] Zero-infinity dilemmas are decision situations that represent apparently small risks of cataclysmic effects, such as risks of a serious accident in the production of nuclear power. The newer environmental problems that I am defining share the characteristics of zero-infinity dilemmas, but have an additional distinguishing feature: The heaviest risks involved in the newer environmental problems fall not upon present actors but upon future generations who will not have participated in the decision to incur the risks; nor will they, in most cases, enjoy the benefits of the currently wasteful activities involved. New environmental problems therefore share all of the difficulties involved in other zero-infinity dilemmas and, in addition, raise serious ethical issues of intergenerational equity. The distinctive character of these emerging problems suggests that it will be useful to designate them as "third-generation problems."

One means to assess the hypothesis of this book, that a policy consensus is emerging among environmentalists, is to test it against third-generation problems. If the concerns environmentalists are just beginning to formulate with regard to newer environmental problems can be comfortably and illuminatingly described within the proposed consensus, then this will lend further credence to my hypothesis. If, on the other hand, environmentalists find no guidance for future policies to address these new problems, which threaten to dominate discussions of environmental policy over coming decades, then this failure will call into question the viability of the proposed consensus, or at least my formulation of it.

In approaching such risks, the Aggregators have often extended the standard benefit-cost approach to decisions involving risk by considering increments of risk as a cost and decrements of risk a benefit.[4] In this manner, economists who accept the standard benefit-cost model can incorporate concerns regarding risks into their decision models because avoidance of risk in the future will have value to decision-makers in the present. Since the decisions we make now must be justified by considerations apparent to us now, rational decision-making must incorporate currently available information and currently accepted ideas about what is valuable. The present-value approach therefore has unquestioned appeal.

Third-generation environmental problems, however, place a heavy burden on certain aspects of the standard analytic approach.[5] Mishan and Page concluded that decisions entailing risks of this magnitude are not susceptible to quantification, and that the decision to accept the risk should rest rather on a reasonable choice between risk acceptance and risk aversion.[6] The argument of Mishan and Page turns mainly on the difficulty of reliably quantifying risk values of events that would entail catastrophic costs. Another feature of third-generation problems that causes difficulties for standard analyses is the long time-latency of the effects in question. The newer problems are extrapolated theoretically according to models that assume that the effects may increase rapidly after decades of minimal effects. For example, the greenhouse effect, a warming of the atmosphere mainly resulting from the consumption of fossil fuels, is projected to cause relatively small changes in temperature in the near future. These early effects would be difficult to discriminate from natural, cyclical changes in climate. Some climatological models predict, however, that the effect will accelerate over time and, by the time the problem is clearly demonstrable, irreversible destabilizations of contexts will occur. These will cause great harm to the next and succeeding generations. It can be argued that losses in biological diversity will have similarly accelerating effects.

Standard resource analyses usually deal with time preference by discounting future benefits and costs. This approach can be quite useful in dealing with relatively short frames of time. Further, economists can cite considerable theoretical and empirical evidence that human decision-makers display a distinct time preference for the present.[7] Discounting the costs and benefits of various policies across time provides an important systematization of this common human practice.

Policy decisions such as those exemplified in newer environmental problems, however, involve effects with time horizons extending far past individual lifetimes and would seem to require understanding of *social*, as well as *individual*, discounting of future effects. As long as risks are felt by the same generation that creates them through incremental choices, the problem of risk assessment and management can plausibly be considered within a consensual model in which risk-aversive activities will have present value to the choice-maker. But when intergenerational times are considered and the risks incurred in the present fall mainly upon future individuals, the model lacks this consensual claim to validity. Whereas consumers are well trained in choosing between their own short- and long-term interests by the school of hard knocks, they must distribute cross-generational risks through altruistic impulses toward the not-yet-born. Therefore, third-generation environmental problems necessarily involve questions regarding the fairness of intergenerational distribution of risk, and it is difficult to see how the standard economic model can adequately conceptualize these essentially ethical considerations.[8]

The contextual approach to environmental management recognizes constraints based in the dynamic interplay between specific actions (such as use of CFCs in aerosols or continued unrestrained use of fossil fuels) and the larger, normally slower-changing context in which those decisions are implemented. According to contextualism, there exists a threshold within which individual decisions to use CFC aerosols or burn fossil fuels will have insignificant impacts on the larger environmental context, which in these cases we can understand as the atmospheric envelope surrounding the earth. The atmosphere has significant resilience and can damp out consequences of the activities in question up to some point; if, however, major and persistent trends over several generations (such as accelerating use of fossil fuels since the onset of the industrial revolution) continue indefinitely, the atmospheric threshold is exceeded, and the autonomous and slow-changing characteristics of the atmosphere can undergo rapid change, such as changes in temperature many times more rapid than those that would normally occur with the advance and recession of the ice ages. According to this reasoning, activities that threaten no thresholds raise no ethical questions—they can be decided freely be individuals on the basis of choice. As scientific models, such as the climatological models now being developed to indicate global warming trends, indicate cross-generational impacts, the contextual approach to environmental management counsels behavioral adjustments, incentives to reverse the trend, mitigative efforts such as massive tree planting, and eventually constraints to avoid rapid destabilization of the large-scale environmental context.

The reductionistic model, which interprets all values as preferences that individuals express in the present, relies exclusively on social-scientific data. What if individuals in the present, either uncaringly or in ignorance of models understood by scientists, act in ways that will destroy the context of economic and cultural values that would have been expressed in the future? To answer this question, it would seem, a rational decision procedure would incorporate a biologically based criterion such as the one developed in Chapter 7.

The usefulness of the contextual approach in conceptualizing newer environmental problems can be more specifically illustrated by examining problems in managing national parks in a period of rapid climatological change caused by the greenhouse effect. Linda Brubaker has argued that

the effects of climatological change on managed vegetative systems must
by discussed with respect to three time scales of variation:

1. Long-term variations (ten thousand to one hundred thousand years)
   are exemplified by changes from glacial to interglacial climates. Responses of vegetation to climate on this scale involve major changes
   in the ranges of species and result in evolutionary changes in the
   genetic composition of the species themselves.
2. Intermediate-term variations (two thousand to five thousand years)
   represent major changes within glacial and interglacial periods and,
   on this scale, species respond mainly through behavioral adjustments
   and perhaps through some evolutionary change.
3. Short-term variations (up to 500 years) may occur within the life-span
   of single individuals of long-lived species such as trees. Established
   plants may undergo physiological or morphological changes in response to variations in temperature and these changes may also affect
   rates of reproduction and establishment of the various species. These
   latter changes may affect the composition of the vegetative community.[9]

On scale (1), the current interglacial period, the holocene, began about
ten thousand years ago and appears to be one of the warmest periods in
the past two million years. By examining the fossil record indicating change
on scale (1), Brubaker concludes that, historically, plant communities have
been transient assemblages, seldom persisting more that two thousand to
five thousand years in the fossil record. This evidence leads Brubaker to
conclude that

> the ice-age vegetation of North America was markedly different from the
> present-day vegetation. Most modern forest types did not exist in the past,
> and many of the most important forest species of today's forests were rare.
> Thus modern communities should not be thought of as highly coevolved
> complexes of species bound together by tightly linked and balanced inter-
> actions. Modern communities have not had long histories, and the species
> rather than the community should be the focus when considering the con-
> sequences of future environmental change. Historical records show that
> species can expand rapidly as climate becomes less limiting; they also sug-
> gest that species that are rare on the modern landscape have potential for
> becoming common under changed climates (and vice versa).[10]

Similarly, Brubaker argues that, since variations on the intermediate scale
(2) also cause major changes in composition of plant communities, the preservation focus again should be on the total complement of species rather
than on attempting to preserve, in a rigid manner, particular complexes of
species current today. Brubaker's argument, which may or may not be correct in detail, illustrates in its general pattern the ecological/contextual ap-

proach to environmental management. Brubaker recommends, in effect, management in three distinct scales of time.

Her reasoning therefore supports the argument, made in Chapter 8, that total diversity should be the major focus of species preservation efforts, and also represents comparatively good news regarding the future of management to protect biological complexity. If we can stabilize the larger ecological context by reducing extreme trends, and if we act aggressively to protect total diversity of large areas, the creative and productive forces of the future may not be destroyed. Specialist species, even those that are confined by natural forces or by human activities to small preserves, will provide a healthy group of competitors to create new assemblages of species that remain efficient in exploiting and transporting energy.

On the shortest scale of time (3), she concludes: "Decade and century-long climatic variations can affect both species ranges and growth rates," and consequently that "present-day forest stands may have established under different conditions than exist today and thus may not return to current composition following disturbance."[11] Environmental management must be based on a dynamic, contextual criterion of health. There must be, according to environmentalists, a forward-looking but ecologically formulated standard by which to judge the future effects of current activities.

Brubaker's analysis therefore requires a dynamic model in which changes in microhabitats takes place against the backdrop of a constantly changing mosaic of macrohabitats; the macrohabitats, in turn, change also against the backdrop of dynamically changing climate. Brubaker's approach therefore suggests a hierarchy of nested systems changing on different spatial and temporal scales.

While Brubaker cautions that several scenarios for climate change are plausible, that the sensitivities to climate of many species are unknown, and therefore that current projections of vegetative changes are too speculative to dictate management plans at this time, the general approach she suggests provides the beginnings of a means to study the management implications of global climate change. In particular, by tracing patterns of vegetational change on scales (1) and (2), projections of future changes on scale (3) can be modeled simply by assuming that accelerated global climate change caused by the build-up of greenhouse gases will be similar to climatological changes on the longer time scales indicated in the fossil record and in pollen deposits in lake sediments. By modeling changes of vegetation patterns within a hierarchy of nested systems changing according to different scales of time, it is possible to create analogues to the rapid changes in vegetative patterns that will result if rapid global warming occurs. It is in this sense that natural history, as well as human history, can clarify policy issues.

A contextual model of management problems may therefore provide important insights to guide environmental management in the face of rapid atmospheric changes brought about by human-induced (and relatively rapid) changes in concentrations of greenhouse gases in the atmosphere. By extension, similar modeling may prove useful in understanding change brought about by acid deposition, by ozone depletion, and by rapid deforestation in the tropics.

The dust bowl proved that rapid and pervasive changes in human use of land can lead to breakdowns of entire geographical systems; such breakdowns both signal and exacerbate destruction of the complexity and integrity of the land system, destroying the complex pathways by which energy flows through the system. Hierarchy theory, which aspires to model and relate the various temporal scales that constitute ecological complexity, may therefore provide a precise means to explore the thresholds and limits beyond which human-induced changes in larger systems such as the whole atmosphere are likely to result in ecological breakdowns with unacceptable consequences.

While this exploration of the impact of third-generation, environmental problems has been of necessity both sketchy and speculative, perhaps enough has been said to suggest some future applications of contextualism in environmental management. First-generation problems were mainly problems in the use of resources—wastefulness in timber production, rapid depletion of nonrenewable resources, or specific threats to a species or a spectacular natural area. Accordingly, these early problems were understood atomistically, and the analysis of them could be largely aggregative. The shift to second-generation environmental problems such as the overuse of persistent pesticides and the spread of pollution, more generally, forced emphasis on the ecological context, and imposed a systematic view on environmentalists. Individual actions take on normative aspects, mainly insofar as they represent larger trends destabilizing ecological systems. Pesticides were seen to affect whole ecological systems; air pollution affects the atmosphere above cities, resulting in smog and unhealthy air; and dumping industrial and household wastes in a stream was recognized as a threat to whole watersheds. For these problems, aggregative techniques were less adequate and environmentalists since Carson have consistently called for limits on economic behaviors that negatively affect larger systems.

The growing prominence of third-generation problems can be interpreted as another step in this progression. Just as second-generation problems were superimposed on first-generation problems (that did not go away), third-generation problems represent effects of changes in intermediate subsystems, such as the atmosphere over cities or tropical forests, on yet larger systems, such as the global climate system. The progression from first-generation to third-generation problems therefore represents a shift from isolated problems to which an atomistic response is not entirely implausible, toward global problems that can be addressed only by paying close attention to larger and larger, and more and more inclusive systems. As newer environmental problems affect dynamic, autonomous processes in larger and larger spatial and temporal contexts, contextual management will become increasingly necessary to understand and manage the interrelated problems that plague the beleaguered biosphere.

The Veil of Intertemporal Ignorance

Reflecting on the fierce green fire he saw dying in the old she-wolf's eyes, Leopold concluded that we must learn to think like a mountain. Leopold flirted with a literal and moral organicism, but was unsure how much reality to ascribe to the organicist analogy: is the mountain *really* alive and thinking? Is the mountain itself a moral being?

These metaphysical and moral questions had little effect on his management ideas, however, because he never questioned that man will be henceforth "the captain of the adventuring ship" and he believed that all of his management strategies could be justified according to a longsighted and "noble" anthropocentrism. As long as it was coupled with the unquestioned truth that all things are interrelated, Leopold believed anthropocentrism would support the land-use policies he advocated in the Southwest Territories.

Looked at across generations, alteration of the American landscape, especially in the arid and semiarid regions, had been a sad story of progressive deterioration and declining land health. Starving deer, deepening gullies, and the dust bowl were the legacy of just a few generations of land "management" by white settlers. One of the strongest elements of Leopold's land ethic is its sensitivity to cross-generational impacts of land management. Thinking like a mountain is thinking in the mountain's longer and slower-changing frame of time—time measured in generations, not seasons.

These long-term relationships must therefore be modeled in intergenerational time. Contextualism, modeled hierarchically, showed promise to exhibit the dynamic relationships between managed cells and their larger temporal and spatial context. Might it also provide a model for understanding problems in intertemporal morality?

Standard theories of intertemporal ethics, or of "intergenerational equity," always have reached the conclusion that our obligations to proximate generations (our children and their children, for example) differ radically from our obligations to distant generations. In an extreme form such theories state that, while we might have substantial obligations to the next generation—our children—we have *no* obligations whatsoever to the generations that succeed them.

After surveying both utilitarian and justice-based theories of our obligations to the future, John Passmore concludes:

> So whether we approach the problem of obligations to posterity by way of [the utilitarians] Bentham and Sidgwick, [or the rights theorists] Rawls or Golding, we are led to something like the same conclusion: Our obligations are to *immediate* posterity, we ought to try to improve the world so that we shall be able to hand it over to our immediate successors in a better condition [than we found it in], and that is all.[12]

While this theory is given broad support among highly respected philosophers, it conflicts with some strong intuitions of environmentalists. Consider three examples. First, take the case of storage of radioactive wastes materials. Environmentalists have insisted on measures that would store radioactive wastes for the duration, measured in millennia, of their toxicity to humans. Second, consider the reactions of environmentalists to global warming models. If it turns out that severe impacts of global warming are unlikely to be felt for two generations, environmentalists would not stop insisting on remedial activity.[13] As a third example, consider environmentalists' widely held belief that the ongoing *processes of nature,* including ecological and evolutionary processes, have great value. But this obligation does not extend just to the next generation. If events that we set in motion today play themselves out in such a way as to destroy the last autonomously functioning systems a century hence, environmentalists believe we will have done wrong. We will have deprived future generations of their aesthetic and natural scientific heritage.

When moral intuitions conflict with moral principles, we sometimes give up our intuitions, provided that the theoretical arguments are strong. Passmore's argument, in both the utilitarian and rights-bases cases, turns crucially on our inability to project human individual wants and interests into the further future: Since we cannot predict what future consumers will want or need, nor what new resources may become available to them through advancing technology, it is senseless to act to protect resources they may neither want nor need.[14] His reasoning is little more than an application of the individualistic bias of contemporary ethics, a bias he assumes as a premise. If Passmore's argument assumes individualism—that causing the deterioration of an *environmental system* can never be a moral issue—the argument will justifiably be rejected by environmentalists as unpersuasive and question-begging. That systems can be made ill, and that human causes of this illness raise moral issues are, after all, the core ideas of the land ethic!

By reversing the direction of the individualistic argument of Passmore, however, we can accept the environmentalists' intuitively felt obligations to the distant future as valid and can conclude, not as Passmore does, "so much the worse for future generations," but "so much the worse for standard, individualistic ethics."[15] If the exclusive concentration on individual preferences, interests, and rights characteristic of theories of utilitarianism and of justice prohibits recognition of felt obligations to distant generations, then those theories are inadequate, by their essential nature, to deal with second- and third-generation environmental problems.

If, following contextualism and hierarchy theory, we incorporate the ecological axioms listed in the third section of Chapter 9 and employ the associated evaluative definition of ecosystem health, we have already placed a value on protecting the creative productivity of biotic systems—as the context that gives meaning to human activity, they carry the value of all human uses and interactions with nature into the indefinite future. Intertemporal contextual thinking therefore presupposes an independent *value* axiom, the Axiom of Future Value: The continuance and thriving of the human species (and its evolutionary successors) is a good thing, and every generation is obliged to do what is necessary to perpetuate that good.[16] The obligation to perpetuate and protect the human species is therefore accepted as a fundamental moral axiom, which exists independently of obligations to individuals. The conservative philosopher Edmund Burke expresses this organicist attitude toward society as "a partnership not only between those who are living, but between those who are living, those who are dead and those who are to be born."[17]

Passmore despairs of understanding our obligations to distant generations because he is looking at the wrong scale. It is true that we cannot criticize the trend toward anthropogenically caused global warming on the basis of harms to specific individuals. If your children, for example, have the foresight to buy land in the northern plains of Canada, your grandchildren may have reason to rejoice at global warming. If, on the other hand, they are stuck with investments in Atlantic City hotels, they may complain. Environmentalists do not wish to argue the intragenerational questions of fairness to individuals, but to argue that current global warming models suggest change that is too rapid to be intergenerationally benign.

The choice to plant wheat or to take the car instead of the bicycle takes on intergenerational implications when a threshold is approached; it raises issues in intertemporal ethics when exemplary of trends that may, given local conditions and the fragility of the larger bioregional system, have serious intergenerational consequences. That threshold is contextual because it depends on local conditions and the fragility of the larger bioregional systems in question. Building on this concept of ecosystem fragility, we can circumscribe a set of environmental problems that have intergenerational implications from those that do not. What we need, in effect, is a moral filter that corresponds to the coarse-grained filter of contextual, hierarchial thinking and that focuses on phenomena in the larger scale of bioregional change, not on particular effects on individuals within that region.

John Rawls, in his infinitely fertile treatise *A Theory of Justice,* suggests such a moral filter, which he calls the "veil of ignorance." Imagine a rational, self-interested individual, Ric, who chooses the general rules for a just society, knowing that he will have to live in that society subsequently. In fact, the veil of ignorance is many veils. By varying Ric's knowledge, we can filter out individually motivated interests based on gender, class, economic status, and so on.

For our present purposes, we can place Ric behind a veil of intergenerational ignorance—he must design a society that he would be willing to live in without knowing the generation in which he is going to live. Now, if Ric accepts Leopold's concern that land use practices and other activities of modern humans, which are distinguished by enormous technological capabilities and growing populations, may alter bioregional systems so rapidly that there will be significant and detrimental impacts on the well-being of society, then he would design a society that constrains trends that destabilize larger environmental contexts.

If it turns out that Ric is born into a primitive society, or even European medieval society, the intergenerational constraints may seem minimal; with hindsight and improved scientific monitoring, however, we can say that conformity to these constraints would have protected the countryside in Greece and China, for example, from the disastrous effect of deforestation and erosion, and might have extended the duration of prior civilizations, and padded their fall when the reason was not ecological destruction. Contextualism understands moral obligations to land systems in a historical context and emphasizes that, given our knowledge of ecological fragility and our powerful technological capabilities to alter those systems, a generation such as ours has special obligations. As Ric foresees a society such as our own, which alters nature rapidly and has available frightening models projecting cataclysmic changes in the environmental context, he would expect us to question the moral acceptability of our violent activities. He would choose a society that would struggle to delineate parameters and thresholds, based on the best models of biology, ecology, climatology, and so on. These parameters and thresholds would, in turn, imply constraints on trends in individual behaviors that threaten to accelerate destabilizing changes in a normally slow-changing environing system. From a moral viewpoint, these constraints would represent "fair" treatment of future generations—the treatment a rational, self-interested chooser would insist upon if he did not know which generation he will inhabit.

When environmentalists accept an obligation to future generations, they do not see this as an obligation to any particular individuals; the relationship occurs at the interface of two systems (human economics, demography, and so forth) with geophysical systems. Viewed on that level, environmentalists believe that there are biological and climatological constraints and that these correspond to moral constraints limiting the extent to which any generation could fairly degrade the world's resources. Believing this, it is not surprising that environmentalists also believe that we are morally required to undertake stabilizing actions when projections show that trends in individual behavior threaten a biological or climatological threshold and institute accelerating changes in the environing systems.

These obligations are viewed holistically, organically—they are owed to the future, just as we are indebted to our forefathers, not individually but collectively, for our cultural heritage; these obligations derive from faith in the value of the human struggle, the Axiom of Future Value. Environmentalists do not wish to meddle in the individual affairs of future generations; they want simply to ensure that those individual dealings take place against a livable environmental context.

Environmentalists' moral intuitions that we should limit fossil fuel use to slow the greenhouse effect, for example, are not based on a balancing of winners and losers, but on the belief that we ought not to destabilize the normally slow-changing systems on which our daily activities depend. That belief depends on (1) a factual scientific model that emphasizes the importance of complexity and autonomously functioning natural systems; (2) a concern that rapid, anthropogenic alterations of larger, environing systems such as the atmosphere will threaten the stability of conditions necessary for an orderly society; and (3) a moral commitment not to engage recklessly in activities that perhaps will cause irreversible changes in the normally slow-changing atmospheric system within which all species, including our own, act to maintain their well-being.

WE HAVE CONCERNED ourselves mainly with examples of preemptive constraints limiting the physical context in which our society will evolve in the future: Will there still be wildness? Will the climate be sufficiently stable for human and multispecies communities to thrive? Will the future be spared costly clean-ups of our toxic wastes? But for environmentalists it is almost as important to recognize the cultural implications of their emerging worldview. Environmentalists have fully endorsed Muir's idea that we should set aside beautiful natural areas because these are essential to the formation of sound human values. They believe that contact with nature is contact with our past, and embodies wisdom. If the moral future of a given regional lifestyle is to have roots in that region's past and if distant offspring are to be expected to respect the present as a part of their past, they must see past generations as sensitive to the ongoing organism that is our common society and to the larger organism that is nature. The lesson of ecology is that one cannot care for the future of the human race without caring for the future of its context. Destruction of our cultural and natural history accelerates the dynamic of moral change.

If we act as individualists and do not value the systematic context of the human values we pass to the next generation, we will have acted out Passmore's self-fulfilling prophecy. If we pay no heed to the context in which future generations form and question their values, they will indeed live in a different world than we do; we will have contributed nothing to their culture; we will be strangers across only two generations. That is the consequence of acting on Passmore's theoretical arguments—if we pay the future no heed, they will pay us no heed. And if we destroy the natural link in our melting-pot community, the common American experience of "pioneering," we have severed our history. Our offspring will no longer understand our aesthetic sensibilities; our culture will be irrelevant to them. Context gives meaning to all experience; consequently, it is a shared context that allows shared meanings—what we call culture—to survive across generations.

Environmentalists' search for a moral vision is the search for a lifestyle and an associated set of ideas that can guide cultures to a harmonious relationship with the land. A land ethic, on this view, is the moral thread that links past, present, and future individuals in a common culture. That culture can be perpetuated only if it respects limits inherent in the land context—for continuity in that land context gives shared meaning to cultures as they unfold through time.

The wilderness, healthy land, as Leopold said so well, is the only moral talisman we have left:

> Ability to see the cultural value of wilderness boils down, in the last analysis, to a question of intellectual humility. The shallow-minded modern who has lost his rootage in the land assumes that he has already discovered what is important; it is such who prate of empires, political or economic, that will last a thousand years. It is only the scholar who appreciates that all history consists of successive excursions from a single starting-point, to which man returns again and again to organize yet another search for a durable scale of values. It is only the scholar who understands why the wilderness gives definition and meaning to the human enterprise.[18]

12

Interspecific Ethics

What makes deep ecology deep? This is perhaps the most perplexing question about the much-discussed but little-understood deep ecology movement. Its spokespersons, who are mostly West Coast and Australian academics, all cite, with some degree of affirmation, Norwegian philosopher Arne Naess's 1974 article, "The Shallow and the Deep Ecology Movement." But nobody, not even Naess himself, still accepts the seven principles of deep ecology that were outlined in the original paper. There seems to be agreement, however, that the movement gains its unity and identity from a shared belief that nature has value independent of its uses for human purposes. To put their point critically, movement proponents all believe that our current environmental policies are in a profound sense "unjust" to other species. Most simply, the deep ecology movement has clearly defined itself *in opposition to* "shallow ecologists," or as some of them put it less pejoratively, "reform environmentalists," who are taken to include all of the mainline environmental groups.

Deep ecology, given its self-proclaimed opposition to all "shallow" approaches, represents a modern version of the idea that environmentalists sort themselves into two broad classifications based on opposed motives. More precisely, we can understand deep ecologists' characterization of two opposition groups as a theory intended to explain the behavior of contemporary environmentalists: Environmentalists pursue two opposed approaches to environmental problems because some believe, while others do not, that elements of nature have independent value. Some environmentalists, according to this theory, are interested only in conserving natural resources for future human use; others, deep ecologists, act to protect nature for its own sake.

If indeed deep ecologists are offering such an explanatory theory, it is important to ask exactly what behavioral phenomena are to be explained: Do reform environmentalists pursue *policies* that differ significantly from those pursued by deep ecologists? Or do they pursue the same policies, but employ importantly different *strategies and tactics* in these pursuits? These two questions will be the subject of the next two sections, respectively. Along the way, we can also assess the strengths and weaknesses of the deep ecologists' contribution to environmental goals.

Deep Policies?

Are the policy objectives espoused by deep ecologists more radical than the announced positions of mainline environmental groups? To give this question some focus, we can limit our attention to policies in the area of species protection and preservation of biological diversity. By asking whether deep ecologists have different policy objectives in this limited area, we can gain insight into the implications of the idea that our current policies are "unjust" to other species and perhaps better understand the contrast between deep and shallow ecology.[1]

Deep ecologists do not, apparently, follow early preservationists who on some occasions adopted an extreme isolationist stance, implying that *all* management by humans is unnecessary and immoral; they do, however, claim that their tendency "is heavily in the direction of minimal human disruption of natural systems," and that they believe in " 'hands-off' techniques."[2] I take this to mean that deep ecologists adopt nonmanagement as an *ideal*, but that they recognize that management is necessary *in practice*, in some cases. As Devall and Sessions say: "It would seem to be compatible with deep ecology principles and 'righteous management' that, in general, if humans have distressed an ecosystem, they have an obligation to help heal that system."[3] But deep ecologists also recognize (sometimes, and without enthusiasm) that humans must also exploit nature to live. Devall and Sessions therefore allow another exception to the ideal: "Nonhuman nature should be used only for *vital* needs."[4]

Mainline environmental groups have likewise enthusiastically supported natural regulation where possible, even while recognizing similar sources of limitation. They too would continue to manage systems that have already been altered in profound ways, as when large predators have been extirpated from wilderness areas and when wetlands have been drained. Recognizing that all systems are affected by human activities to some degree, mainline environmentalists conclude that they should nevertheless manage to protect the integrity of ecological systems—their autonomous function-ing, which is supported by their ecological complexity. This commitment to complexity also encourages environmentalists to propose restorative ex-periments, such as wolf reintroductions and dechannelization of rivers. With respect to ideals, and the exception allowing restoration, mainline environ-mentalists and deep ecologists apparently agree.

The second exception, as formulated by Devall and Sessions, is more problematic. Mainline environmentalists certainly accept uses of nature broader than for vital needs, since they have followed Muir in strongly supporting nature parks, hiking trails, and so forth. It is difficult to believe that Devall and Sessions interpret "vital needs" narrowly and literally. Do they really oppose all outdoor recreation? Do they oppose using timber to print books and to build libraries and museums? If so, they differ from mainline environmentalists, but one doubts that they will find many follow-ers in their ascetic lifestyle.[5]

Warwick Fox, a talented young Australian philosopher, has recently un-dertaken a sympathetic but critical examination of deep ecology. Fox takes deep ecologists to task for "employing a definition of anthropocentrism which is so overly exclusive that it condemns more or less *any* theory of value that attempts to guide 'realistic praxis'." Deep ecologists, he says, have "shied away from considering situations of genuine value conflict and [have] not come forth with ethical guidelines for those situations where some form of killing, exploitation or suppression *is* necessitated."[6] As academics, the spokespersons for deep ecology have been able to avoid adopting policies on difficult, real-world cases such as elk destroying their wolf-free ranges, feral goats destroying indigenous vegetation on fragile islands, or park fa-cilities overwhelmed by human visitors. We are left with the task of deter-mining, by implication, what policies would follow from the abstract moral principle that defines the movement.

Perhaps the intrepid Ric can again help us, this time to design a fair society of mixed species. To filter out any "species-ist" notions Ric may be afflicted with, we assume Ric will design a mixed society he will live within, and that Ric can see through the veil just well enough to tell that there will be one rational and very powerful species, but he does not know whether he will be a member of that species or some other. Ric's deliberations may help us to understand moral relationships on the synchronic, outward-look-ing scale as they did on the diachronic, forward-looking scale, by filtering out the selfish interests Ric might unfairly protect if he knew he would exist as a human.[7]

At first Ric will no doubt be taken aback by the assumption that he may, after acting as an honored philosopher king on one side of the veil of ig-norance, turn out on the reality side of the veil to be a cockroach or a

Furbish lousewort. Taken aback but not daunted, Ric strikes into un-charted philosophical territory and devises a righteous management plan based on the moral principles of deep ecology.

But immediately Ric faces a crucial ambiguity. Must he design a man-agement plan that is fair to all *members* of all species? Or should he devise a plan whereby every species, *as a species*, is given equal consideration? In the former case, Ric would pay attention to every human activity having an impact on any living individual; in the latter case Ric would pay atten-tion to human activities when they constitute a trend that accelerates changes in the larger, environmental context in which all species live, making the niches of some species less tenable at a rapid rate.[8] Individual members of other species would thereby be given special protection only if they were members of increasingly rare or endangered species.

Fox, to his credit, recognizes that deep ecologists' reticence to specify precise policies stems from this basic ambiguity in the statement of their central principle, which virtually equates two distinct ideas—the moral idea of biospherical equality and a metaphysical principle of self-realization as a part of a larger whole. The first idea expresses disgust at human arrogance and suggests we should treat all species as we do other humans; metaphys-ical holism originates in ecology and recognizes that no individual can be understood independently of its environment. As Frithjof Capra puts this point, nature is "a complicated web of relations between the various parts of the whole," implying that the individual is constituted by its relation-ships to its environment.[9] As this point is sometimes put, "The earth is my body."

If it seems odd that two principles so apparently different have been conflated, deep ecologists' reasoning is straightforward: The metaphysical principle of self-as-part-of-a-whole is taken as basic. It transforms the self into a world without boundaries and implies that, when I harm the whole (or any part of it), I harm my (extended) self. Therefore, I should treat all parts of nature as I would myself. This extension of the Golden Rule, plus holism, implies biospherical egalitarianism; I should treat every other in-dividual in nature as I would treat myself.[10]

If Ric wishes to operationalize the individualistic, egalitarian aspect of this two-headed principle as a plan for righteous management, he must design a plan that is *fair to all individuals of all species*. But this, it turns out, is much easier to say than to do in cases of real conflict of interest. Consider the three cases mentioned above (p. 222)—herd culling, removal of human-introduced exotics, and shortages of hiking trails. It can never be "fair" by human standards to kill 10 percent of an elk population because it exceeds the capacity of its range. "Justified" killing of humans requires that they commit a capital offence but elk are as morally innocent as human babies. They cannot recognize the consequences of their breeding and browsing, nor can they voluntarily abstain from these activities. The argument is similar for goats. Finally, since humans can live without hiking trails, deep ecologists apparently oppose cutting new hiking trails. This latter consequence is especially ironic as an application of Muir's idea of righteous management; he crusaded to make natural wonders accessible to citizens because he thought this exposure would make them allies in preservation.

Recognizing that unqualified egalitarianism will lead to paralysis in situations of unavoidable conflict, some deep ecologists propose a weaker *practical* principle to supplement their *theoretical* principle. Fox, for example, follows Birch and Cobb in ascribing differential intrinsic value to organisms depending on their complexity and capacity for richness of experience.[11] While this prompted a reprimand from Sessions for introducing a "pecking order into the moral barnyard," it does allow deep ecologists to placate animal liberationists and advocate Alan Watts's explanation that he was a vegetarian "because cows scream louder than carrots."[12]

This relaxation may allow deep ecologists to avoid starvation without offending their colleagues the animal liberationists. But what about plant liberationists? Worse, the weaker principle will not help with the difficult cases of herd culling and removal of exotics. Elk and goats, presumably, scream louder than the plants they eat. Nevertheless, Sessions has a point. One suspects that "complexity and capacity for richness" will map onto the phylogenetic scale in a predictably anthropocentric pattern.

The plucky Ric will now see that the problem with the deep ecologists' principle as a theory of righteous management is not that it is too *strict* but that it is inappropriately *individualistic*. The deep ecologists assume that a new morality for the environment will merely extend an individualistic morality to all species. But this simple extensionism conflicts with their holistic metaphysic, which denies that any individual can be seen *simply* as an individual. The 120,000th elk cannot be treated equally with one of the last California condors—not, at least, on a reasonable *environmental* ethic.

Ethical extensionists cannot understand our obligation to other species for the same reason that intergenerational ethicists cannot understand our obligation to distant generations—they are looking on the *individual* level. If Ric tries to be fair to all individuals of all species, taken either in the strong sense of Devall and Sessions or in the weaker sense of Fox, he will face infinite and irresoluble conflicts. The only way to make all interactions "fair" in the sense of human ethics would be to interfere daily in the lives of other species, regulating the kills of predators. Ric might even be driven eventually to follow Stephen Sapontzis, who seriously proposes a world purged of all predation and pacifically composed of plants and herbivores.[13] It is a simple fact of nature that wild species live in a world of conflict. We can make that world "fair" only by taming it, and that would be the greatest interference of all.

But Ric would remember that deep ecologists stand for *minimal management,* and minimal management presupposes that each individual should be seen as a part of a larger, systematic, and autonomously functioning whole. That whole can be maintained only if each individual realizes itself as part of a larger context in which there are inevitable "injustices" to individuals of all species. Ric will therefore content himself with a criterion of righteous management based on *interspecific impartiality,* rather than individual fairness. He will apply a systematic, demographic criterion to each species. For example, humans are morally justified in culling elk only if the humans are willing, similarly and impartially, to limit their own populations when they exceed their carrying capacity.[14] This coarse-grained version of righteous management, which would focus on individuals only when their species was in trouble, or only when their populations were exploding, would square with the American naturalist tradition that puts (roughly) equal value on saving any species, and can be understood scientifically as an application of ecological management, which understands and evaluates populations functionally within a larger ecological context. This approach would monitor demographics of all species, instituting recovery plans for any species when it became rare over a short period of time, and reducing populations of species that explode unnaturally.

An effort would be made to protect at least representative wild ecosystems as stable habitat for all species but, as long as populations of wild species are changing slowly in response to changes in ecological time, they would be left alone. In this sense, humans are accepting an ecological role, as competitors who alter their own ecological context, even as they alter those of other species. Conscious decisions augmented by awesome technology lead to rapid changes in those impacts; therefore, conscious understanding of our competitors' role implies that we should moderate our impacts because the context experienced by other species is the same context experienced by us.

On the questions of subsistence exploitation and sport hunting, environmentalists disagree. But they can still act for the common-denominator objective of saving ecological complexity by focusing their biodiversity policy on *systematic* rather than individual issues. Note that this coarse-grained, demographic principle of impartiality is simply the Safe Minimum Standard of Conservation approach—the policy of protecting species and ecological systems provided the costs are bearable—that is favored by mainline environmental groups.[15] Deep ecologists' policy on biological diversity would differ from other environmentalists only if they interpret their task in inappropriately individualistic terms.

Ric can draw an important generalization from these deliberations. Once he abandons the impossible goal of designing a management plan that is fair to all individuals, he can altogether abandon the awkward subterfuge of interspecific ignorance. We noted in the last section that Ric, if ignorant of the generation he will inhabit, would impose rules on each human generation for the protection of all species, and of wild, autonomous processes to the extent possible. He would insist that any generation of humans that is growing so fast, in population and in economic and technological development, that it threatens to extinguish one fourth of all species in a generation or so should act to reverse its violent impacts on environing systems. This would imply that these generations, such as our own, should monitor the populations of every species, protect individual members of species whose populations are becoming rare more rapidly than can be explained by ecological forces, and develop positive plans to protect biodiversity on all levels.

It implies, also, that humans have a moral obligation to halt their own population growth as rapidly as possible and perhaps even to reduce populations to ecologically sound levels.

Ric, acting as ignorant of his species, would choose the same policy, even though that policy might allow the regrettable loss of a species if resources are unavailable to save every one. Suppose, for example, that someone suggests that the costs of saving the Furbish lousewort (of which there are few specimens and which seems to support no other species) are too high— that it would be better to channel resources into protecting wetland communities. Ric, choosing as a potential human of unknown generation, will regret this loss because it will reduce diversity and redundancy, and because there is a remote chance that he will exist in a later generation that will be significantly impoverished because of the lack of louseworts. Ric, choosing as a potential organism of unknown species, will regret the decision because he might have turned out to be a lousewort or dependent on louseworts. But once it is given that some species must go, he would make the same choice, regardless of which veil of ignorance he stands behind. Not knowing his generation, it is best to save all species, even at considerable cost; not knowing his species, ditto.

This conclusion suggests, if generalized, that introducing the idea that other species have intrinsic value, that humans should be "fair" to all other species, provides no operationally recognizable constraints on human behavior that are not already implicit in the generalized, cross-temporal obligations to protect a healthy, complex, and autonomously functioning system for the benefit of future generations of humans. Deep ecologists, who cluster around the principle that nature has independent value, should therefore not differ from longsighted anthropocentrists in their policy goals for the protection of biological diversity.

When we question, then, the deep ecologists' hypothesis that environmental groups pursue opposed policies because only some of them rally behind the idea that nature has intrinsic value, we must conclude that there exists no such sharp distinction in policies pursued. We must look elsewhere if we are to understand what makes deep ecology deep.

A Strategy of Depth?

Perhaps the disdain of deep ecologists for other environmentalists, their claim to have achieved a deeper level, has more to do with strategies and tactics than with policy goals. Devall and Sessions begin their important overview of the deep ecology movement by noting that reform environmentalists "will easily be labelled as 'just another special interest group.' In order to play the game of politics, they will be required to compromise on every piece of legislation in which they are interested."[16] They approvingly quote Murray Bookchin, who, writing in the social anarchist tradition, asks whether environmentalism will "be marked by a dismal retreat into ideological obscurantism and a mainstream politics that acquires power and effectiveness by following the very stream it should be seeking to divert?"[17] Deep ecologists, following Bookchin, therefore prefer the "minority tradition of direct action."

But again, what seems like a healthy one-on-one disagreement turns out to be a tangled polygon. There is, it turns out, more than one way to be an outsider. Alston Chase, covering a conference on "International Green Movements and the Prospects for a New Environmental/Industrial Politics in the U.S.," notes with irony that the German Greens, who originally built on the ideas of deep ecology but who have given them a Marxist-materialist and distinctly left-wing flavor, now have little in common with American deep ecologists. Kirk Sale, an American deep ecologist at the conference, was almost shouted down as he talked about "living in place" and "bioregionalism." Chase quotes the response of Devall, who also attended the conference: "It does not resonate with me," he said of the German Greens: "They don't see things through the deep-ecology paradigm."[18]

American and Australian deep ecologists are not political revolutionaries; they represent an individualistic and spiritualist movement, seeking an "inner" transformation of persons rather than a mere change in political forms— a new consciousness, not a new government. And, while some activists who flirt with violence, such as the Earth First group, cite deep ecology as their guiding light, the more academic spokespersons for the movement hold such activities at arms length, arguing for "direct actions" but being careful not to endorse violence. They quote Thoreau, not Marx: "Let your life be a friction against the machine." As Devall and Sessions say: "The deep ecology movement involves working on ourselves, what poet-philosopher Gary Snyder calls 'the real work,' the work of really looking at ourselves, of becoming more real."[19] They work from outside the system not as political revolutionaries do, but as spiritual dissenters to all forms of highly organized activities. They are as opposed to communism as to capitalism; both are seen as statist systems that will ruin the earth.[20]

Chase sums up the incompatibility, evident at the International Conference, between the German Greens and the American deep ecologists: "The Germans sought political power; they were into compromise, wheeling and dealing. They were a kind of German Sierra Club. And most important, the German Greens were materialists, while the essence of deep ecology was its spirituality."[21]

American deep ecologists oppose coercion and all "hierarchical and centralized" authority, and this seems to be their major quarrel with mainstream environmentalists. Devall and Sessions question the trend among mainline environmental groups toward "professional leadership in centralized organizations," which they fear will "cut at the roots of volunteerism and could reinforce the trend toward ruling organizations by experts and bureaucracy, and the trend to mass politics."[22] It sounds like a rerun of Muir the uncompromising amateur vs. Pinchot the conniving politician.

But the amateur/professional distinction is here given a new twist: Whereas Muir was an effective, if reluctant, organizer and used the Sierra Club to lead a national lobbying effort that held up the Hetch Hetchy project for over a decade, the deep ecologists wish to place Muir's Sierra Club and all the other major organizations on the professional side of the divide.

Deep ecologists recognize, on one level, the importance of these day-to-day activities of the large national organizations: "This work is valuable," Devall and Sessions concede. "The building of dams, for example, can be stopped by using economic arguments." But they always follow such concessions immediately with qualifiers—"However this approach has certain costs." These costs are explained in two quite different ways, so it becomes necessary to distinguish two variants of the deep ecologists' argument that mainstream environmentalists have gone astray in their strategies and tactics. One version of this argument, which we can call "tactical," criticizes the mainstream groups' emphasis on accepted political approaches, those groups' willingness to work through normal political channels—legislatures, courts, and executive agencies—to lobby and bring suits, to present information at hearings, and sometimes to compromise with the powerful economic forces supporting current exploitative practices. But deep ecologists also criticize mainstream groups for failing to emphasize that nature has independent value. This second, "strategic," line of attack criticizes the mainstream for using human-oriented arguments to support their policy goals. This strategic criticism, then, raises what is in fact a question of values or motivations—deep ecologists believe that employing anthropocentric arguments will in the long run undermine environmentalists' effectiveness in protecting the environment.

Note that, if there were important differences between deep ecological policy goals and humanly oriented ones, these two arguments would be difficult to separate—deep ecologists could be seen as criticizing mainstream groups for pursuing the wrong policies *because* they have the wrong ideology. If, however, as was argued in the last section, there exist no important differences between deep and reform policy goals, it becomes important to examine these two arguments separately; we must consider both of the following, distinct arguments, one tactical and one strategic, against mainstream environmentalists: (1) the *political* tactics of the mainstream groups will fail to achieve the common-denominator political objectives of all environmentalists (tactics of political compromise should be eschewed in favor of direct action and consciousness-raising); and (2) the mainstreamers will not succeed in their *educational* goals of improving the perceptions, attitudes, values, and behaviors of the general population because their political strategies sometimes appeal to anthropocentric values.

1. The first objection seems to be that the more the large groups take on the bureaucracies and the forces of development, the more they look like those bureaucracies, and the more they share their tactics. Hence, "reformist activists often feel trapped in the very political system they criticize."[23] What is unclear is why Devall and Sessions try to define themselves in opposition to the tactical approaches of mainline environmental groups. Why not see the approaches as complementary, as did Leopold, who worked simultaneously to improve management practices within the system, even while writing nature essays to "work on" the attitudes, values, and worldview of citizens? If deep and mainstream policy goals are roughly similar, it seems clear that the political tactics of improving policies in the short run while improving perceptions and attitudes in the long run are complementary. For example, it can be argued, historically, that the Supreme Court case *Brown v. the Board of Education*, exercised as much influence on public opinion regarding race relations as vice versa. Surely the environmental movement will fail to achieve its general goals unless it

succeeds *both* in reforming the current system politically *and* in the educational task of changing the consumerist attitudes currently dominant in American society.

The deep ecologists' insistence that working within the system has costs as well as benefits is nevertheless well taken. An exclusive commitment to insider tactics and polite compromise with opponents can, in fact, leave environmentalists impotent. For example, Michael McCloskey, Chairman of the Sierra Club, has recently concluded that working within the regulatory procedures of the Environmental Protection Agency has proved unproductive in some cases, citing the case of Alar, a carcinogenic additive used to color and preserve apples. After years of working within the agency and the courts to reduce the use of Alar, the Natural Resources Defense Council went public, charging EPA with foot-dragging because of coziness between regulators within the agency and industrial interests. The more direct, public appeal was successful in terminating the use of Alar in a few days, accomplishing what could not be done through years of lobbying the EPA. McCloskey proposed that more direct appeals to the public, perhaps even "radical" direct actions, may prove to be more successful in some cases than insider lobbying.[24]

From a more theoretical viewpoint, however, the tactical question of working within the system versus employing direct actions to raise consciousness, improve public understanding, and achieve bold results has nothing to do with the question of whether nature has independent value. Anthropocentrists can become as impatient as biocentrists with the geologic pace of reform within the system. Bookchin, who deep ecologists cite as a guiding light regarding political tactics of direct action, writes from the anthropocentric tradition of anarchism and has recently lambasted deep ecologists, faulting the movement for its concentration on nonhuman values and for its inability to draw important distinctions between destructive and nondestructive human impacts.[25] If there is a sharp distinction between environmentalists regarding whether to work within the system or through direct action—which seems doubtful given environmentalists' pragmatic tendency to choose whichever political tactic seems most likely to work in a given situation—that distinction has nothing to do with their philosophical commitment to anthropocentrism or to biocentrism.

2. The second critique, the charge by deep ecologists that mainstream environmentalists are too quick to use anthropocentric arguments, *sounds*, at first, like a major disagreement about ultimate values, and it may, indeed, rest on such a disagreement. But judged within the context of the shared view of all environmentalists, including deep ecologists, that the public must be educated and must adopt a less materialistic and more systematic view of ecological systems, it *does* become a reasonable question

of general strategies. Should environmentalists adopt a single worldview, a single value orientation, in all of their activities? Understood in this way, the deep ecologists would be arguing that, because mainstreamers must often use anthropocentric arguments to be successful in the political arena, they will be ineffective as leaders in the longer-term task of reforming the consumerist attitudes and values of Americans. Assuming a common goal of reforming American attitudes and values, one can indeed ask whether such reform is more likely to be achieved if environmentalists adopt a single, "orthodox" worldview and value position or if they remain moral pluralists.

It is certainly not the case that one can question consumerism only from a perspective that values nature independently; the rich tradition of Judaeo-Christian stewardship, though staunchly anthropocentric, has a lot to say, for example, about materialism and greed. And remember also that Muir, in all of his public writings, castigated the "Servants of Mammon" in Christian rhetoric, carefully avoiding references to pantheism or mysticism. Contemporary environmentalists, in the tradition of Muir, Leopold, and Carson, have similarly used *both* anthropocentric and biocentric arguments. The deep ecologists' strategic argument is therefore best understood as a warning that, in the long run, promiscuity regarding ultimate principles—as opposed to a "pure" commitment to the biocentric principle—will undermine environmentalists' goals of improving human treatment of nonhuman elements of nature.

Deep ecologists apparently believe, contrary to this long-standing pluralistic attitude, that purity of purpose and an unswerving commitment to biocentric principles will in the long run lead to a better outcome for nature than will a mixed morality of political pragmatism. An uncompromising commitment to biocentrism, they think, will lead to an uncompromising policy stance and in the end will increase the chances of a Final Victory for the forces of Good over the forces of Evil. As an empirical hypothesis about which strategies are more likely to achieve given goals, this proposal is no doubt deserving of careful consideration. In so considering, however, it is important not to confuse motivational and moral questions with strategic ones. In the next section, we will examine, philosophically, the truth of the claim that nature has value independent of humans; here, with the focus on strategic questions, the issue is whether environmentalists' pluralistic stance causes them to compromise in situations where they should not and whether it weakens their moral appeal in the attempt to change the attitudes and values of (admittedly overconsumptive) Americans.

It is on this strategic point that I believe deep ecologists make their most important contribution. It is, indeed, important to ask whether, in the long run, environmentalists would improve their chances of success if they, following Thoreau, emphasize moral self-examination or if, following Pinchot, they work to build political consensus. But it is important, also, to consider the inclusive option, the strategy of Leopold, which was to pursue two, parallel projects—to work on the long-term goal of reforming arrogant and environmentally insensitive attitudes even while working within the political system to minimize damage from current practices. Understood in this way, deep ecologists are arguing that the tactics of political compromise weaken the movement morally, that a consistent adherence to biocentric arguments would cause environmentalists to push harder on humans and increase the likelihood of a more utopian result. It is, without question, important for mainstream environmentalists to consider this possibility.

My concern is that deep ecologists, under the influence of moral monism, seem to formulate this important strategic question in exclusivist terms. Because they think that anthropocentrism excludes biocentrism and vice versa, they do not seriously consider Leopold's inclusive, dual strategy; assuming that the interests of humans and of nonhuman species are essentially at odds, they often insist that all actions should be based on a single moral principle, that of biocentrism. One fears that underlying this assumption that we must in all cases choose is a largely unexamined assumption—a corollary of monism—that human interests and nonhuman interests are necessarily in conflict.

More recently, however, deep ecologists have taken a less unyielding stance on the question of whether nature has human-independent value. Fox, for example, in his more recent work has dissociated himself not only from extensionism, but from any "values-in-nature" position at all. He now doubts that deep ecologists wish to present an alternative "axiology" (which we can read as "a complete and well-connected value component of a worldview") at all. They wish merely to destroy anthropocentrism. Fox quotes Naess, who now emphasizes that biocentric egalitarianism is "a statement of non-anthropocentrism." It is, he says, "not a formal axiological position."[26]

Naess admits that "the abstract term 'biospherical egalitarianism in principle' and certain similar terms which I have used, do perhaps more harm than good. They suggest a positive *doctrine*, and that is too much."[27] Correcting his earlier close association of self-realization with a strong, positive view of our obligations not to harm other organisms because of their human-independent value, Fox now follows Naess in asserting only self-realization as a positive philosophical position, relegating references to independent value in nature to the role of rhetorical devices that "should be taken colloquially rather than technically," because they are "deliberately framed in everyday, non-philosophical language."

This emendation results in a significant reorientation of the deep ecology viewpoint. The goals of deep ecology are not to establish a positive moral principle, but to broaden perception. Naess says: "I'm not so much interested in ethics or morals. I'm interested in how we experience the world. . . . If you experience the world so and so then you don't kill. If you articulate your experience then it can be a philosophy of religion."[28] Sessions agrees: "The search, then, as I understand it, is not for environmental ethics but for ecological consciousness." These clarifications are very helpful; they make it clear that the deep ecologists' central concern is to raise the "deeper" questions of how we perceive and value nature, to emphasize that environmentalists ought not to become so involved in playing political games that they lose sight of the central insight, prominent in Muir and Leopold, that humans must struggle to overcome their arrogance. Once deep ecologists back off from an insistence that the new and less arrogant approach to values must take a single form, a positive commitment to independent values in nature, it becomes clear that they are advocating asking "deep questions," emphasizing the importance of consciousness-raising, and calling for a comprehensive reevaluation of our society's goals and values.

On this point, deep ecologists are clearly in step with the mainline environmental groups, except that the latter are more likely to pursue a dual strategy, while deep ecologists prefer only to "work on themselves." At this point, then, the original exclusivism of the deep ecologists, the basis for a *distinction* between deep ecologists and shallow ecologists/reform environmentalists has dissolved. If deep ecologists are merely pointing out that there are two tasks, changing policies and changing minds, and that they prefer to emphasize the latter task, why insist that there exists a dichotomy among environmental*ists*? Their point, once purged of a doctrinaire commitment to independent values in nature, amounts to a commitment, on their part, to pursue one approach to the exclusion of the other, and a commitment to the belief that, if environmentalists pursue *only* a pragmatic, political strategy, they will not succeed in their long-term goals. It may well be that mainline environmental groups must, occasionally, be reminded of this point—but it would be a terrible distortion of historical and contemporary fact to suggest that mainline environmental groups deny the importance of working on changing attitudes and values. At most this criticism involves a reminder to groups deeply involved in day-to-day political battles that they must not forget to step back from the fray and ask deep questions, as well.

A Deeper Commitment?

The strongest contribution of the deep ecologists, as we have seen, is their single-minded commitment to the proposition that environmentalists must retain their moral focus, that they will not succeed in their long-term goals by being nothing more than another interest group practicing the art of the possible. That point is granted; we have seen that, in many contexts, environmentalists have found it necessary to appeal to moral values, human and otherwise. The remaining question is whether environmentalists should adopt a single theory of value such as biocentric egalitarianism to govern their actions in all contexts, or whether they should be moral pluralists.

Given that Naess and other important deep ecologists have given up a claim to any specific axiomatic system of value, why not take the next step, repudiating the exclusivism that attends the distinction between depth and shallowness, and join with other environmentalists in a common quest for a criterion of acceptable behavior in our activities affecting nature? One suspects that, behind the deep ecologists' reluctance to abandon exclusivism based on biocentrism is an assumption that that principle will stiffen environmentalists' spine, make them unyielding in the face of pressure from economic interest groups, and strengthen their voice against materialism, consumerism, and short-sighted greed. A commitment to biocentric values set against human values, it is apparently assumed, explains why deep ecologists can credit themselves with a willingness to press harder against human frailty and give nature its due.

To evaluate this final claim, that embracing biocentrism will deepen environmentalists' commitment, we must examine more philosophically the exact meaning of the hypothesis that nature has value independent of humans. There are, it turns out, two importantly different senses in which nature can be said to have independent value. First, values in nature can be independent of humans, *absolutely*. Applying this *strong* conception of human-independence, which we can call "intrinsic" value, biocentrism implies that nature had value before *Homo sapiens* evolved, and will continue to have this value even after the human species blows itself up or fades into extinction. Intrinsic value can exist in the absence of any conscious valuers, at least of the finite sort.

Other theorists defend a less heroic version of the theory, that nature has "inherent" value, meaning nature has value that does not depend upon human *values*. On this theory, all value is experienced by some human, but some human valuations are understood *noninstrumentally*.[29] To use Calicott's analogy, parents value their children, not just instrumentally, but also inherently, for their own sake.[30] Here there is no question that independent value exists within human consciousness; Callicott refers to inherent value so understood as intrinsic value in "a truncated sense," and asserts that it "retains only half of its traditional meaning." It "is valuable *for* its own sake, *for* itself, but it is not valuable in itself, that is, completely independently of any consciousness."[31] It is this version that expresses Leopold's belief that the land ethic will be culturally based.

So, we can identify three very distinct theories regarding the value of nature:

1. *Anthropocentrism:* All value in nonhuman nature is instrumental value, dependent upon contributions to some human value;
2. *Inherentism:* All value in nonhuman nature is dependent on human consciousness, but some of this value is not derivative upon human values; and
3. *Intrinsicalism:* some value in nature is independent of human values *and* human consciousness.

The difference between intrinsic value and inherent value is epistemological, not moral. Intrinsic value, being prior to human conceptualization, is therefore *discovered*, while inherent value is *posited within* a human theory or worldview. Intrinsic value, existing prior to any worldview, enforces itself on any adequate conceptualization of the world; inherent value, on the other hand, is attributed, by humans, as a part of their conceptualization, description, and valuing of the world. This represents a crucial distinction, epistemologically, because intrinsic value, unlike inherent value, cannot be supported by scientific or any other cultural resources—it must be supported independent of all experience. It must be "intuitively" true.

Roderick Nash interprets the history of environmental ethics as the gradual realization that the rights of man, firmly grounded in the strong, a priori sense of intrinsic value, should apply to all entities. Nash therefore cites John Locke and Thomas Jefferson as important precursors of environmental ethics. Locke and Jefferson had the correct *theory* of independent value, but they "trembled before the veil," and arbitrarily limited rights to human freemen. Deep ecologists and other forward-looking environmental ethicists, Nash thinks, have had their moral vision cleared by the civil rights movement and by the animal rights movement and are about to correct the temerity of Locke and Jefferson by recognizing that all nature has the same rights that we twentieth-century Western persons do.

The suggestion that environmentalists who adopt biocentrism will be more committed and less compromising no doubt gains comfort from this interpretation of independent value: Commitment to self-evident rights in nature would provide environmentalists with an objectively discoverable, logically supportable, and culturally independent moral talisman. They rest their moral case ultimately on certain and self-evident truths.

The disadvantage of this strong interpretation is that it rests on the very Cartesian dualism that deep ecologists wish to scuttle. While John Locke is justly credited with undermining the Cartesian conception of science based on self-evidence, he never questioned intuitionism in ethics. Locke and Jefferson were deists—they believed that God created the world according to laws of nature but did not interfere in day-to-day operations of the planet. This somewhat awkward dualism of two worlds allowed them to see the physical world as determined and amenable to scientific explanation, even while the pure light of natural reason ruled in the realm of ethics. The rights of man could be proclaimed as self-evident, even as rationalism crumbled in the world of science, because enlightenment thinkers clung to the belief in self-evident moral principles supported by rational intuition.

This vestigal Cartesianism in ethics corresponds to the distinction between intrinsic value on the one hand and inherent and instrumental value on the other. If the exclusivism associated with moral toughness rests, as I have argued it does, on this epistemological priority of self-evidence, exclusivism is acquired at a high price. Besides encumbering environmentalists with a whole set of questionable philosophical assumptions, and encouraging the confusions about individual rights of members of wild species, this intrinsicalist approach to independent value also dulls their analytic tools.

As John Dewey argued so perceptively, the a priori style of reasoning pays little attention to particular experiences of nature, because it is independent of all such experience. It assumes there is a single, unitary truth discoverable by philosophical reasoning, and hardly pays attention to particularities. It is better at drawing grand distinctions, such as anthropocentrism versus biocentric egalitarianism, but not so good at smaller distinctions such as justified versus unjustified herd culling. The intuitionism and self-evident reasoning associated with intrinsicalism therefore cuts against the strong commitment of deep ecologists to localism, regionalism, and encouragement of home-grown ideas—all such ideas must be held accountable to universal intuitions, and this has an inevitable homogenizing effect.

Inherentism, on the other hand, follows Leopold's idea that the land ethic will be constructed within a culture, drawing on all the religious, scientific, and moral resources of that culture, not within an abstract world of pure reason. Unharnessed from the idea of self-evident moral rights, inherentism is also more flexible, and can locate moral value in systems as well as individuals. But inherentists, if they truncate the theory of value to scuttle conceptual commitments to dualism, must also truncate any claims to moral single-mindedness based on self-evidence and culturally independent norms. Inherent value is attributed within cultures, experienced within a historical and intellectual context, not *ex nihilo*.

The theory that environmentalists should be sorted into two camps according to commitment, or lack thereof, to the principle that nature has independent value had led us to no important policy differences between environmentalists and their critics among deep ecologists. And, while some environmentalists appear more politically radical in some situations than others, these tactical differences did not correspond to the line drawn by belief in biocentric principles. If it cannot explain moral steadfastness without appeal to Cartesian intuition, perhaps the whole idea that there are deep and shallow ecologists reduces to a distinction desperately seeking a difference. Perhaps the idea of independent value in nature should fall to Occam's razor. But here I simply quibble. Only philosophers are likely, driven by the urge to find unitary principles, to care about this question of monism versus pluralism. To the extent that the environmentalists' dilemma rests on the drive toward moral monism, the search for a single moral principle, it is more a philosophers' dilemma than one for active environmentalists.

The attack on human arrogance, which was mounted as a response to anthropocentrism, was well motivated but badly directed. One need not posit interests contrary to human ones in order to recognize our finitude. If the target is arrogance, a scientifically informed contextualism that sees us as one animal species existing derivatively, even parasitically, as a part of a larger, awesomely wonderful whole should cut us down to size.

The Convergence Hypothesis

Despite the popularity of moral monism among philosophers, moral chaos is not an inevitable consequence of moral pluralism. Moral chaos can be avoided in a pluralistic system of values, provided there are clear *rules of application* for deciding which principles apply in each given situation, and clear *priority rules* for deciding which principles take precedence when more than one applies. If these conditions hold for a pluralistic system of values, we can say that it is *integrated*. An integrated, pluralistic system will not be subject to either manipulation or arbitrary application, provided it is followed in good faith.

What, then, lends impetus to moral monism at the expense of an integrated pluralism? As W. V. O. Quine said in a metaphysical context, monism represents a "taste for desert landscapes." If it is merely a taste, it should be noted that it was acquired very early in Western philosophy. Combining metaphysical aridity with hydrological abundance, Thales said, around 600 B.C., "All is water." Judging by standards of simplicity alone, we've not improved on Thales' view. As pluralists, however, we can argue that, whatever the charms of theoretical elegance, other criteria, such as adequacy to the complexities of reality, may justify a more complex set of principles. If this pluralist argument carries the day, we should lower our sights, requiring not monism, but some form of integrated pluralism.

Environmentalists appeal to elements of at least seven distinct worldviews and a diversity of associated values. Nevertheless, we have been able to trace an emerging consensus regarding desirable environmental policies. That considerable consensus represents an integration of a variety of values. Why not try for an integration of human and nonhuman interests rather than positing human-independent values in contradistinction to human ones?

Philosophers' concern about multiple moral principles is not without basis. I learned this as a child when I finally caught on to my older brother's scam: he stuck me with doing the breakfast dishes three days running by varying the principle between "You got them out, you wash them" (if I had eaten first) and "You finished last, so you should do the dishes" (if I had a late breakfast). While this will not rank with great moral abuses in history, it illustrates the philosopher's point: Multiple moral principles, applied willy-nilly or, worse, manipulated cleverly, can result in injustice. These problems can be swept elegantly away, philosophers have realized, by assuming that all activities will be governed by a single, over-arching moral principle. But the breakfast example has a moral bite only because my brother and I shared a loathing of dishwashing and hence had clearly contrary interests. Similarly, moral monism becomes a pressing issue only on the assumption that the interests of nonhuman nature *are* in conflict with the interests of the human species. The urgency with which deep ecologists urge a recognition of human-independent values to act as a counterpoise against human self-interest rests firmly on this, the motivating assumption of monists and exclusivists. But is this assumption true? Suppose for the moment that it is not.

Environmentalists should prefer a multilevel integrated system because environmental problems are so often problems of integration. Making a profit in farming is a good. Avoiding erosion is a good. Protecting wildlife is a good. A monistic system insists that a single criterion should apply to all of these goods; an integrated system, however, admits all of these values are goods within a particular subsystem. Normally, the contextualist can concentrate on maximizing goods appropriate to the subsystem under consideration; at the same time, it recognizes that all subsystems also participate in a larger context and balances must be struck.

Contextual thinking encourages us to focus on environmental problems as involving impacts on multiple levels, in different scales, and it may occur that the same action can have differential moral value, depending on the context in which it occurs. Our Farmer Jones, who is deciding whether to clear his woodlot and plant wheat there, is acting within several nested systems simultaneously. On one level, he is acting within a free enterprise social and economic system, within which he will decide mainly in terms of his individual interests. Within a larger, synchronous system, Jones's action may be catastrophic to individual members of wild species, and widespread habitat destruction may raise serious moral concerns.

But the decision can also be evaluated for its impacts on the environment, or context. If Jones's woodlot is on a steep slope, or in a watershed that is prone to serious erosion of topsoil, the decision may have impacts on future uses of the land. On this level, Jones's decision can be seen as an event on the interface of a cultural and a natural system. If Farmer Jones's decision is part of a major trend, woodlot-clearing may trigger both accelerating changes in context and preemptive constraints. Since the future possibilities of individuals in the social system may depend on Jones's decision, it should be examined for impacts on this larger system, which can be expected to unfold in a longer scale of time.

A hierarchical system of value therefore opens the door to new possibilities for understanding environmental ethics. Environmentalists need not choose between the worldview of anthropocentric economic reductionism and biocentrism. Another possibility is an hierarchically organized and *integrated* system of values. Such a system aspires not to reduction to a single scale of value, but to second-level principles that explain and justify the proper realm, or system of application, of a variety of constraints associated with cultural and biological limits inherent in larger, regional systems.

Environmentalists need a unifying vision for the future. Contextualism can provide an integrative map for the consideration of environmental problems as they unfold outward into larger and larger systems, provided this structure is sufficient to support the common-denominator objectives shared by environmentalists of all stripes. I believe that an integrated system, a worldview that is unified by a commitment to a variety of values on various levels, holds more promise for unifying environmentalism than either of the monistic and exclusivist value approaches of economic or biocentric reductionists.

The key to the integrative approach we have developed is a recognition that the vision of environmentalism is unified, not by a shared commitment to a single value, but by a shared belief in scientific naturalism and its associated belief that all things in nature are related in complex, hierarchically organized systems—the commitment to scientific naturalism, its associated natural aesthetic, and to a belief that humans evolve their personalities and cultures within environing systems that are, ultimately, shaped and limited by hierarchical constraints. Conservation biology is not a value-free science—its basic concept is ecosystem "health." But ecosystem health can be understood only in a cultural context—the land ethic is a locally determined sense of the good life, constructed with a careful and loving eye on the natural constraints imposed by the ecological and climatological context of that life.

Because of a shared commitment to scientific naturalism, environmentalists of differing value commitments gravitate toward similar policies, because they believe that a scientific understanding of ecological systems determines the available means to pursue their diverse goals. No long-term human values can be protected without protecting the context in which they evolved. Scientific naturalism thereby enforces upon all environmentalists a basically holistic, nonatomistic approach to environmental problems.

Although they are fascinated with the disagreement raging over the center, or centers, of environmental values, active environmentalists have made their peace over this issue by accepting an empirical hypothesis—the convergence hypothesis. Environmentalists believe that policies serving the interests of the human species as a whole, and in the long run, will serve also the "interests" of nature, and vice versa. When David Brower says, "Everything I have done as an environmentalist can be justified in human terms,"[32] a philosopher who assumes anthropocentrism and biocentrism are in conflict will conclude that Brower has no concern for other species. But this is manifestly untrue. Like so many great environmentalists before him, he is relying on the convergence hypothesis.

While empirical in nature, the convergence hypothesis is not a precisely formulated hypothesis open to direct test in a series of dramatic experiments. Although cases where human interests and nonhuman interests seem to conflict are clearly relevant to the case, environmentalists will not surrender it easily. The convergence hypothesis therefore has a dual status as (1) a very general empirical hypothesis and (2) an article of environmentalists' faith. In this respect it plays a role similar to the medieval belief in the geocentric universe. It is an empirical hypothesis, but it is not to be given up at the drop of an epicycle.[33] Similarly, advocates of the convergence hypothesis, when faced with an example in which the interests of another species seem in conflict with long-term human interests, will dismiss it, claiming the example has not yet been viewed in sufficiently long temporal terms.

The convergence hypothesis functions, then, as an item of faith, guiding environmentalists' ongoing search for a rational solution to environmental problems. Since environmentalists believe that the hypothesis follows as a corollary of the systematic emphasis of contextualism, it is supported more by scientific theory than by particular observations. The convergence hypothesis rests firmly on the central insight of ecology—that all things in nature are interrelated. If humans damage the larger context shared by both humans and other species there will, eventually, be negative impacts on all species, humans included. But the hypothesis is also informed by another law of ecology, the law of complexity, which tells us that things are not equally related.

This approach is neither monistic—it posits no single moral principle determining morality in all subsystems—nor aggregative—it does not sum results across systems. It is hierarchical—it applies to each moral problem those locally determined principles determined by a careful look at the local and regional context shaping that problem. This approach integrates man into the ecological system—it avoids isolationism by recognizing that human cultures have, since time immemorial, shaped their context. Also, it avoids atomism, and tries for a broader integration of social values, including wilderness values and a hope for a future that respects and shares our values.

Ecological science, hierarchical thinking, and the convergence hypothesis work together to define the limits of individual behavior. But they do so not by meddling in every individual decision, but by encouraging local freedom and determination. Environmental constraints apply only when limits are approached. And a society with wisdom and foresight may be able to encourage enough diversity in voluntary land use decisions and careful management of publicly owned lands to require little interference with individual freedoms.

Because individual humans will affect natural systems as predators and competitors, as well as symbionts, their activities will not always be in the interest of individual members of wild species. Viewed thus individually, the convergence hypothesis is obviously false. Interpreted as applying on a larger scale, the level at which cultures affect the natural systems they inhabit, the convergence hypothesis mimics Leopold's land ethic, implying that (1) humans must develop a sense of moral obligation to the protection of land systems, including their complexity, internal organization, and the energy pathways they embody; and (2) this obligation can be viewed as an obligation *either* to the land system itself *or* to the collective system that is our culture's distant future. The domain of the convergence hypothesis, then, is the larger scale within which cultures affect functioning natural systems.[34]

Viewed hierarchically, in fact, our moral relations with other living creatures begins to look more manageable. Animal liberationists have been severely criticized for failing to note the apparent differences between our moral relations with domestic animals and those with wild animals. For example, failing to feed one's domestic livestock is simple cruelty, whereas the decision not to feed wild animals who have overbrowsed their winter range may be the considered judgment of an environmental manager, one that is agonized over and finally accepted as the best means to reduce the herd. Alternatively, any of us would come to the rescue of a pet bird if it were under siege by a predator, tame or wild, but many of us would not feel such an obligation to halt predation of a wild bird by a fox. We might even stand back, in awe, and watch the spectacle. If there is only one principle governing our treatment of other animals, and that principle is that we should do all in our power to reduce their pain, then we should remove wild animals from their natural habitats where they are threatened daily by pain and death, and place them in zoos where they can be fed and protected.[35]

Monistic reductionism holds little promise to unify environmentalists behind a single banner, because it cannot comprehend the variety of values and worldviews that appeal to environmentalists on these various levels. Contextual thinking encourages us to value actions differently depending on the context in which they are analyzed. This scientific model provides the general structure for an integrated system of value, with different values taking precedence in different management contexts. This approach to environmental ethics has the considerable advantage that it localizes and regionalizes the environmental ethic. A search for such an ethic would not, then, take place in philosophy journals, but in public forums, such as the Chesapeake Bay Regional Council. The policies such groups set, after input from philosophers, scientists, and managers and after public decision and debate, would determine the "ethic" for that area. If these discussions are successful, they will open an unending public discussion of the environmental goals for the region. If that ethic is inadequate, the result is cultural suicide. The culture of an area can be passed on only if the autonomously functioning context that gives meaning to that culture is preserved.

Contextualism helps us to explain the intuition that our obligations to wild and domestic animals are different. The mixed natural/social context in which we encounter wild animals is different from the context in which we encounter domestic animals. In the latter case we have entered an implicit contract with domestic animals; we owe them protection as reciprocation for the benefits they bestow on us in a community of *intermingled* species. We have limited their free behavior patterns by breeding and conditioning; we thereby accept a responsibility toward their protection.

Environmentalists insist that the context in which we encounter wild animals is different—it is mixed, but should not be intermingled. Within a mixed context of people and wild species, humans should forbear from interfering in the lives of individual wild animals; interactions should be watched, demographics of wild populations should be monitored, and recovery plans should be instituted if populations decline rapidly in response to human-induced changes in the larger context.

Convergence, unlike reduction, goes both directions. Sometimes, rapidly declining populations of wild species indicate rapid degeneration of the context. This canary-in-the-mine idea, so effectively developed by Rachel Carson in *Silent Spring*, shows how other species can be used as indicators that the human environment is becoming less favorable. But the convergence hypothesis works the other way, as well. Many environmentalists believe that we owe to future generations relatively intact tropical rainforests—and we can discharge that duty only if we act decisively to protect populations of beleaguered species.

The convergence hypothesis is the hopeful hypothesis. If, in fact, there is no obligation to protect the context of human activities, or if the human species exists in fundamental opposition to the healthiness of ecological systems, there is no escape from the environmentalists' dilemma. Either we will be economic Aggregators and incrementally sell our cultural and natural heritage down the proverbial river, or we will embrace biospecies egalitarianism and be cursed with a choice between self-starvation and constant sin against that moral law. Contextualism and pluralism, however, hold out a reasonable hope for constructing a culturally *and* biologically determined conception of the good life—one recognizing that every species modifies its context, but also recognizing that the scales on which those modifications take place will determine the quality of all future life.

Epilogue:
Differing Senses of Place

Perceiving in Place

There is a new storm brewing over Chesapeake Bay fisheries; only four years after the moratorium on taking the popular rockfish was imposed, catches of the region's newly favored sportfishing species, the bluefish, dropped precipitously in 1989. Conservationists, worried about the effects of shoreline pollution and rapacious harvesting of the once plentiful bluefish, are beginning to talk about controls on fishermen. Likening the present situation to the slaughter of the bison or the passenger pigeon, they note that bluefish can be caught in large numbers (especially with sonar assistance in locating schools), and amateur fishermen who have no use for a dozen large fish often take them home only to throw them stinking and uncleaned in the trash a few days later. Harley Speir, a fisheries expert at the Maryland Department of Natural Resources supports a campaign to voluntarily release bluefish once caught: "If you ain't going to eat it, don't kill it. There certainly aren't enough to go around if everybody takes as many as they want."[1]

Hearings are being held on proposed first-ever limits on the catch of bluefish. The battle lines are forming; still angered by the rockfish ban, charter captains deny that the bluefish is overfished and are preparing to fight any restrictions on bluefish catches. They argue that the bad 1989 season was due to hard luck. They believe that there are not fewer bluefish, but that the fish have stayed out of the bay because heavy rains have reduced water salinity. Environmentalists are less sanguine: "I don't think

244

it's safe to just dismiss what's happening as weather related," says William Goldsborough of the Chesapeake Bay Foundation. "We've got enough information to be on the alert that this could be a real reduction in the bluefish population."[2]

This emerging situation should by now seem strikingly familiar. Individuals, who have exploited a resource for years, gathering the fruits of an apparently inexhaustible source of nature's bounty, develop an economic interest and an attitude that they have an individual or proprietary "right" to its continued free use. In the area of land use, to cite another example from the Chesapeake region, the Critical Areas Commission has recognized the need for strong local zoning to limit development in the immediate vicinity of the bay and its tributaries. Local governments are asked to develop local ordinances that protect critical areas. On the local level, developers and landowners are fighting to implement this system of controls so as to maximize their own development rights to the fullest.[3]

The activities of charter captains and of real estate developers have been modeled illuminatingly by Garrett Hardin as individual actions affecting a "commons."[4] Individuals act to expand their own "herd," or in the case of bluefish, their own "catch," knowing they will benefit fully from their take. If those actions threaten the larger, resource-producing system, resulting costs will be shared by the entire society, they explicitly or implicitly reason. Individuals acting without restrictions to exploit a common resource are thus driven inexorably by the logic of the situation toward yet another "tragedy" in the unfolding history of resource use.

Hardin's model shows that environmental problems have a common structure, and that this structure is inherent in any situation in which individuals, acting in their individual self-interest, exploit a common social resource.[5] Hierarchy theory and the contextual approach to resource management expand Hardin's insight by positing a larger and slow-scaled environing system that changes dynamically but more or less autonomously. Rapid change, such as a rapid decline in bluefish stocks or water quality are conceptualized as illnesses in the larger system.

According to this model, it is in the nature of environmental problems that they eventually emerge on a higher, contextual level as the activities of individuals, such as charter captains, fishermen, and farmers, tend toward more and more intense use. The model therefore explains the insight, recurring throughout this book, that environmentalists are usually reactive toward social and environmental trends. It is in the nature of the environmentalists' response to these problems that they occupy a more synoptic place; in trying to think like a mountain, environmentalists understand human activities as they affect larger systems, in their larger spatial and temporal contexts.

We have shown that, however interesting and important the philosophical question of whether nonhuman elements of nature have intrinsic value, answers to this question do not correspond in any direct way to important disagreements regarding environmental objectives and policies. Long-sighted anthropocentrists and ecocentrists tend to adopt more and more similar policies as scientific evidence is gathered, because both value systems—and several others as well—point toward the common-denominator objective of protecting ecological contexts. Environmentalists, of course, will continue to disagree about what should be done in particular situations. The payoff of this book should be reflected in an improved understanding of these remaining disagreements.

Given this characterization of environmental problems, we can expect that future disputes will center on the proper "place" from which environmental problems should be observed, discussed, and decided. I will use this concept of "place" in an informal, intuitive sense, as when one says to a friend or associate: "I know we disagree, but if you were standing in my place, I think you'd agree with me," and also in a formal sense, as modelled by hierarchy theory. Hierarchy theorists recognize that the analysis of any complex system will depend upon the observer's viewpoint and on the scale of resolution adopted, which is a function of the place the observer assumes in viewing the complex system.[6] Choosing a "place" involves deciding on both (1) the perspective of the resource manager; and (2) the *scale* on which to conceptualize the larger system or context of which it is a part, which involves setting ecosystem boundaries.

Standing in the place of charter captains, who think in the time-scale of economic reasoning and see their personal investment and livelihood threatened, limits on bluefish catches appear as an unjustified attempt by centralized authorities to limit personal freedom. The captain doubts that the meager data represent anything more than a blip in the natural cycles of weather and migration. Conservation-oriented resource managers, from their expanded place, see warning signals in the decline in bluefish harvested. Looking at the problem not just on an annual scale, but also projecting catches into the next decade and beyond, the conservationist suggests voluntary behavioral modifications and perhaps limits on daily take.

From the environmentalists' vantage point, on the larger systematic and contextual level, the decline in bluefish in the bay is serious. Environmentalists worry that it may be a replay of the rockfish case. If the brief-frame decisions of individual fishermen are reducing the bluefish populations rapidly, the fishery could be destroyed before data determining the limits of exploitation are gathered. Reliable data to document that overfishing is causing the decline of the bluefish fishery are very difficult and slow to gather, because they regard a slow-changing system. It is always difficult, in such cases, to distinguish true trends from cyclical changes.

The problem is that the captains and fishermen may well be correct: Next year may bring record harvests of bluefish. Nobody, least of all environmentalists, doubts the existence of natural cycles. If environmentalists cry "wolf" in a case that turns out to reflect only a natural cycle, however, they can lose their credibility. It is in the nature of their position that they are always waiting for data to support their claims that accelerated changes in contextual systems are imminent. They are often, therefore, reduced to saying, "Better safe than sorry."

In the fishermen's frame of time, the environmentalist is proposing limitations on their individual freedoms. Looked at in the frame of time of environmentalists who advocate "thinking like a bay," this means simply paying attention to the larger and slower-scaled context, the scale of time in which populations wax and wane, the context of the fishermen's behavior. Looked at cross-generationally, fishing restrictions seem like a minor limitation on individual (short-term) freedoms in the interest of protecting those very freedoms through intergenerational time. Environmentalists, by arguing that we should approach limits of nature cautiously, are trying to hold options—freedoms—open to future generations, but in doing so, they inevitably find themselves trying to restrict those same individual freedoms in a shorter frame of time.

Environmental problems of the future will unfold, I am predicting, in this arena of multiple levels; environmentalists will battle their opponents, and also each other, regarding the proper place from which to understand and solve environmental problems. These political differences also express themselves in disagreements regarding the proper political processes for addressing resource issues. As R. Neil Sampson, Executive Vice-President of the American Forestry Association says:

> We've put so much faith in impact statements and our ability to do everything on a rational basis that we're now in a situation where, if a group doesn't like a decision that a Forest Supervisor makes on the Clearwater National Forest in Idaho, they can have that decision out of Idaho and in Washington just like that—and out of the Forest Service and into Congress or the courts just like that, and at no cost.
>
> But it'd be surprising how people would change their views if you could pack them up and take them out there and see the local conditions.[7]

As Sampson's comments show, the sense of place in environmental decisions has a political aspect. We saw the same phenomenon in Chapter 7, when the debate between balancers and moralists in pollution control manifested itself as a battle over whether Congress or the EPA will determine

standards for compliance with targets in pollution control. Environmentalists feel more comfortable dealing with Congress rather than the EPA because members of Congress are sensitive to many constituencies and varied
values. The messy situation on the Hill provides environmentalists with a
battlefield on which they can use their varied worldviews and their political
clout.

Choice of a "place" from which to diagnose, discuss, and solve environmental problems, however, is not just political; it also has conceptual, scientific, and moral aspects. For example, if pollution standards are set at
the EPA, they will be discussed in the reductionistic language of welfare
economics; the relevant "facts" will concern human preferences for automobiles and for clean air, and the values will be individually conceived.
Choosing a place, therefore, is inseparable from choosing a worldview. For
this reason, discussions of environmental policy often sound like exercises
in turf protection.

This assessment can be expressed more formally: The structure of environmental problems, which often occur at the interface of human and cultural systems and their natural contexts, entails that environmentalists will
view environmental problems from a more synoptic viewpoint than will
their opponents. They will see, or try to see, environmental problems on
the horizon of public policy, problems that will occur when current trends
in individual behaviors press against or exceed the parameters of permissible disturbance to larger, environing systems, whether river basin or the
atmosphere. Environmentalists often, therefore, invoke constraints on present individual actions in an effort to protect some broader public good and
therefore express moral constraints on individual action. But the value protected by those constraints will be redeemed in the future, sometimes the
very distant future.

As population grows in a region and available technologies become more
powerful and effective in managing natural systems, environmentalists expect that limits on individual behaviors will have to be more stringent.
Environmentalists stand strongly in favor of population control, but they
recognize that, even in best-case scenarios, rapid growth will continue for
many decades. As Leopold said, greater population entails greater violence
to the land. Protecting beleaguered systems will therefore require more
active management of the cultural/natural interface, as population in an
area increases.

Accordingly, the implications of contextual management will be different
for systems that are more or less "tame." This was noted, in Chapter 9, in
the progression from Yellowstone, through the Pacific Northwest, and on
to the Chesapeake. On one extreme is wilderness, the ultimate test of our
culture: Can we halt the dash toward civilization, and save nature itself,

the wildness that is the origin of our existence, at least in a few places? At the opposite end of the progression are metropolitan areas. Cities, the ultimate context of most future peoples, have a natural context also. Contextualists do not advocate managing Manhattan as they do Yellowstone, to maximize the autonomy of natural forces. And yet environmentalists believe that the management of metropolitan areas in the future must be based upon a sensitive recognition of the effects of cities on their larger natural contexts. The unifying idea of contextualism is, on a more basic level, its systematic feature. It is in the nature of the environmentalists' case that they direct attention to the ways in which trends in individual behaviors affect larger systems. Environmentalists, to successfully make their case, must educate the public to see the overall dynamic and the importance of seeing environmental problems from a synoptic, contextual perspective.

Fragile Freedoms

Back to the beach, one more time: When I saw the little girl with so many sand dollars, I was struck speechless because the languages readily available to an environmentalist were inadequate. Once it is admitted that sand dollars can be exchanged for nickels, the language of economic aggregation encourages the application of a maximization criterion. On that language, the little girl's utilitarian logic was unassailable: More is better. But the traditional language of morality, developed and honed over centuries and millennia to articulate rules for interpersonal behavior among human individuals, was equally inadequate: An extension of the language of individual rights and interests to apply to this interspecific situation would encourage a total prohibition on exploitation—and thus would deny the obvious fact that humans must sometimes exploit elements of nature in order to live and enjoy.

Our search for a way between the horns of the environmentalists' dilemma has led to an emphasis on the *context* of human actions. The family's strip-mining operation, on this view, is wrong primarily because it was inappropriate to its context. The exploitative activity turned the beach into the first stage of a trinket factory; building sandcastles and learning about nature had lost out to an economic perspective. That little remnant of beach was saved for little girls, but for little girls to learn to love and respect their natural context, not for them to learn to exploit its products.

I wish now that I had used the incident as an opportunity (with her parents' consent, of course) to teach the little girl some ecology and natural history. I'll bet I could have interested her in the way that sand dollars make a living. I could turn over a sand dollar so that the little girl could see and feel the kneading of the hundreds of little sucker-feet by which sand dollars pull themselves through the sand while passing some of the particles through their bodies, digesting diatoms from the particles as they pass through and are then flushed out.

I'll bet the little girl would have been fascinated to see that the sand dollar has a pentagonal structure analogous to our head, arms, and legs, but that the sand dollar's nervous system is undifferentiated. Therefore the behavioral repertoire of sand dollars is far more limited than our own. Sand dollars' life in predator-rich lagoons encouraged them to invest in external armour rather than mobility.

This approach, turning the beach into a natural laboratory rather than a trinket factory, is in keeping with the environmentalists' long-standing commitment to the educability of the American public. They believe that if enough people adopt the ecological viewpoint, their approach to environmental policy will win out and their common-denominator goals will be achieved. The natural history approach to the situation on the beach is, in other words, to follow Thoreau, Muir, and Leopold in putting faith in the power of observation and experience to transform worldviews. Here, it is possible to say, is the single greatest failure of the environmental movement. While groups have been quite successful in educating their own members through slick membership magazines, they have made less headway in educating the general population. For example, few schools teach conservation in any systematic way, and most science texts do no more than mention conservation in passing.

But I should not, as part of my lesson, insist that the little girl value the sand dollars *in their own right*. That would be like taking the little girl to a symphony concert and trying to teach her to value one note or to an art museum and trying to teach her to value one brush stroke. We must value nature from our point of view *in a total context*, which includes our cultural history and our natural history. Nature must be valued, from the ecological-evolutionary viewpoint of environmentalists, in its full contemporary complexity and in its largest temporal dynamic.

And this crucial lesson of our dependence on the larger systems of nature can be learned from sand dollars—for sand dollars, just like humans, act within an ecological context. The success of their activities depends on a relatively stable context to which they have adapted. The freedom and creativity of sand dollars is a *constrained* freedom, freedom to adapt to a limiting context. We differ from sand dollars in having a repertoire of behaviors almost infinitely more complex than theirs. But our freedom and creativity is no less than theirs a constrained freedom.

The freedom to collect sand dollars, to catch rockfish and bluefish, and to propel ourselves about the countryside by burning petroleum are all fragile freedoms. They are freedoms that depend on the relatively stable environmental context in which they have evolved. If I could, then, have used the incident on the beach to teach the little girl that sand dollars embody an ancient wisdom from which we can learn, and also to illustrate for her the way in which our activities—just like the activities of sand dollars—are possible, and gain meaning and value, only in a larger context, I would have progressed a good way toward the goal of getting the little girl to put most of the sand dollars back. The strip-mining activities of the family were not wrong in the absolute terms of interpersonal morality; they were inappropriate on a beach set aside for relaxation and enjoyment of nature.

The family's reaction, upon finding sand dollars in the lagoon, was to treat them as an economic resource. But the power boat gave me a clue that they did not really need the nickels, and the little girl's dogged efforts convinced me that she was the loser on the beach. Trips to the beach should be explorations of a larger world than the limited sphere of economic activity in which the little girl will no doubt spend most of the rest of her life. Like Muir and Leopold I should have emphasized the ecstatic aspect of observation and natural history studies. I could have avoided the environmentalists' dilemma by encouraging the little girl to see the world through a lens larger than a cash register. Then, she might have killed some sand dollars to study them, but she would still have *respected* sand dollars as living things with a story to tell. I hope she would also have realized, then, that sand dollars are more valuable alive than dead.

Moralists among environmental ethicists have erred in looking for a value in living things that is *independent* of human valuing. They have therefore forgotten a most elementary point about valuing anything. Valuing always occurs from the viewpoint of a conscious valuer. Since I doubt that sand dollars are conscious, I doubt they are loci of value-expression. To recognize that only the humans are valuing agents at the beach, however, need not enforce the conclusion that the sand dollar will be valued only from the narrow perspective of human economics. If the little girl can learn to value sand dollars in a larger perspective, an ecological context in which sand dollars are fellow travelers in a huge, creative adventure, she will have taken the first tentative steps toward thinking like a lagoon.

Charter captains see restrictions on the taking of bluefish as unjustified infringements of their freedoms. That, as Leopold recognized, represents a failure of *perception*, not value. The captains, used to apparently unlimited bounty from nature, are unable to think like the bay. Environmentalism will succeed if it educates the public so that all citizens are capable of seeing environmental problems at the interface of two systems—the slow-changing systems of nature that change in ecological and evolutionary time and the relatively fast-changing systems of human economics. To the extent that individual freedoms to take bluefish or rockfish depend on the complex, usually slow-changing, systems of nature, they are fragile freedoms. They depend upon, and gain meaning and value within the larger, natural context in which they are pursued.

Does the fragility of freedoms to use nature entail onerous restrictions? Have we, after all, arrived at the depressing conclusion that the future—if environmentalists are correct in seeing the world contextually—will be one of increasing constraints on individual freedoms? Will we, in a world of growing populations and increasing scarcity, be driven to ever more oppressive restrictions on individual freedoms?

I think not. This depressing conclusion follows only if we accept the contextualist worldview incompletely. It is not arrogant to value things from one's own conscious perspective, and to that extent a degree of anthropocentrism is a foregone conclusion. The hard questions will concern which actual activities will be discouraged or limited, and in those arguments environmentalists, anthropocentrists, and ecocentrists alike will support a broadly scientific, ecological, and contextual viewpoint. Most values, from the ecological perspective, depend on saving the ecological systems that are the context of human cultural and economic activities.

Most environmental ethicists have, to date, assumed that we must, to escape arrogance, posit value as independent of human valuing or human valuers. This value has proven to offer little guidance in action and has raised innumerable and intractable questions in the metaphysics of morals. The moral premise of this book is that there exist limits on human treatment of nature—thereby rejecting the implication that humans may do anything they please to their natural context—while leaving the original idea of anthropocentrism—that all value will be perceived from the viewpoint of conscious beings—intact.

To accept the contextualist viewpoint encouraged by an ecological worldview is to recognize that the creative force is outside us. We do not create either energy or biomass—we are derivative beings, who value and choose within a complicated system to which we adapt. Even as we learn more and more about the cunning and creativity of nature, we learn simultaneously that we are finite beings who are free only in the sense that we are free to react creatively and differentially to ever-changing situations outside us, a system that extends beyond our bodies in both time and space.

As Muir and the ecological thinkers who succeeded him emphasized our role as a part of a larger whole, they did not *introduce* the idea that our freedoms are fragile—that idea is clearly implied in the story of the fall from grace and the expulsion from the Garden of Eden. Freedom has always been understood as occurring within constraints. The new idea that must guide environmental ethics is not that our freedoms are limited, but that an important element of those limits exists in nature itself, not in the commandments of a disembodied God or the rights of our fellow humans. The rules governing our treatment of nature are guided neither by the authority of God nor by a priori, precultural moral norms such as rights of natural objects. Environmentalists have been forced to recognize that we must struggle to articulate limits on acceptable behavior by learning more and more about how we affect, and are affected by, our environmental context. The land ethic is nothing more than the latter half of our culture's search for a good life in a good environment. Ecology, the transformative science, prefers an organic metaphor; environmentalists believe that the organic analogy of nature as creative—which is illustrated throughout natural history studies—provides a better metaphor for understanding our adaptive role in larger environments.

In this sense, environmentalists must reject the arrogance involved in the suggestion that humans can do as they please with regard to nature. They must also admit that the creativity of nature is the Great Mystery. If we destroy without understanding, we commit the greatest arrogance, for it is understanding the sweep of millennia, the ability of nature to create more and more elaborate life forms from the deaths of countless individuals, that will ultimately explain our own existence, our correct place. The search for the self-moved mover has only since Darwin become a question in biology. But it was the prophetic Muir who recognized most clearly that science and theology would eventually merge once again as they did in Genesis I. The linchpin of the modern environmental movement is the belief that the study of nature has this ecstatic aspect; the ability to inspire wonder at our "partness" and at the whole of which we are a part is simply the ability to inspire a shift to a new perspective in which nature is an object of contemplation, not exploitation. If that change were to occur, environmentalists believe, there would be more support for contextually sensitive policies.

But still we are left with the disturbing question: How onerous must the restrictions be on future human activities? The answer to this question, I think, must be "It depends." It depends on how we conceive those restrictions. If we, as at present, conceive nature as a machine capable of producing unlimited amounts of a small number of economically useful items, we will view nature and our opportunities statically. If charter captains, who once could offer their clients unlimited catches of bluefish, insist on that freedom indefinitely, they will destroy that freedom. Bluefish catches will eventually be limited by rules and regulations or by natural declines in bluefish stocks. The outlook for human freedom from this static viewpoint looks bleak.

But consider an analogy. Whaling is in the process of fading away as an economically feasible activity. Whale stocks are so depleted that the search becomes ever more expensive. Technology has found substitutes for all but the most esoteric uses of whale oil. Environmentalists therefore insist that whaling is no longer an appropriate activity, even if there are governments that will prop up the dying industry with economic subsidies. If environmentalists and others succeed in the desperate effort to save populations of the great whales, however, there will be a whole new, nonconsumptive, and dynamic industry, whale watching, that will take its place. Children of future generations will pay, it can be assumed, to watch a great whale swim playfully under their boat and breach a few yards away. The fragile freedom to kill whales will be replaced with a more secure freedom, a freedom consonant with the life history of these great, but not reproductively prolific, creatures.

And this suggests the proper answer to charter boat captains who are justifiably wary of catch restrictions given the present attitudes of charter renters. The charter captains have an obligation to educate as well as profit from their customers. The whale-watching case suggests how salable a natural spectacle is. Participation in a bluefish run should be reward enough— and it would be if fishermen carried away information and understanding as well as a couple of bluefish. Charter captains should teach themselves some marine ecology and pass it on as a part of their explanations of why, next year, we're going to release all bluefish but three per fisherman. This is the proper response to a demand for bluefish that cannot be met indefinitely: educate the public and have them pay for it as part of the skills of a competent charter captain. If charter captains will not educate their fishermen, who will?

The writer Annie Dillard expresses a sense of awe, a dogged persistence in seeing and wondering at it all. Nowhere is this sense of wonder—which was, after all, what was missing from the little girl's afternoon at the beach— stronger than in her graphic description of nature's profligacy, a chapter called "Fecundity," in *Pilgrim at Tinker Creek*. She describes aphids, which lay a million eggs to achieve a few adults.[8]

Dillard is saying something very profound, and it holds the key to being content with fragile freedoms. The recognition that our freedoms are fragile is, in a sense, no more than an admission of our own finitude. That, by way of bad news, is nothing new. But Dillard's illustration of the incredible, virtually infinite creativity of nature is the good news. Our freedom and creativity may appear limited when looked at from a conservative viewpoint that insists on pressing fragile freedoms, such as the freedom to take bluefish, rockfish, or whales to their limit. But the same insight should encourage us to recognize that we have the ability to learn, through science, the limitations of populations of rockfish and whales to reproduce, and to encourage alternative, more adaptive human behaviors before an element of nature's productive fabric is destroyed.

And here we see the potentially true nobility of the human species. Unlike the other forces of nature, which react unconsciously to their surroundings, mainly through the weeding out of unfit individuals, we are conscious beings who can adapt consciously to our changing environment. If we can progress beyond the environmentalists' dilemma, which encourages us to understand and value nature either in the limited context of human economics or in the limited context of human ethics, and value nature from *our* point of view, but in its full and glorious context, there is yet hope for the human species.

Conservation biology must now move rapidly to propose a positive criterion of ecological health for natural systems, a criterion that places value neither on simply exploiting the atomistic elements of nature nor on isolating nature and separating human activities from it. But conservationists should not propose their criterion imperiously; they should work with nature interpreters in parks, on television, and in books to involve the public in the search for a land ethic. It should be a positive criterion for the ecological health of systems that recognizes that some human activities are compatible with the ongoing health of the energy pyramid, and others will positively enhance it. The good news for us and future generations of humans is that we, as conscious beings with scientific tools, can occupy a synoptic viewpoint from which we can understand, and protect, the incredible creativity of nature. Our success will depend on how quickly we develop such a positive criterion and how quickly this criterion can become a basis for private actions and public policies. And that, of course, will depend on how successful environmentalists are in encouraging contextual thinking and educating the public in the ecological, systematic viewpoint on nature. It is the firm commitment to the dynamic aspect of human valuing—its reactivity to changing situations—that marks the ecological/evolutionary perspective: humans must understand and value from a realistic perspective by recognizing their role in the larger ecological context.

Notes

Chapter 1

1. Cf. David Ehrenfeld, "The Conservationist's Dilemma," in *The Arrogance of Humanism* (New York: Oxford University Press, 1978). I have modified Ehrenfeld's terminology because, in the historical sections of this book, it has been helpful to use the term "conservationist" more narrowly, to represent only one faction of the early environmental movement.
2. Gifford Pinchot, *Breaking New Ground* (Washington, D.C.: Island Press, 1987; originally published 1947), p. 261.
3. John Muir, *A Thousand-Mile Walk to the Gulf* (Boston: Houghton Mifflin, 1981; originally published 1916), pp. 98–99.
4. Roderick Nash, *Wilderness and the American Mind,* 3rd ed. (New Haven: Yale University Press, 1982), p. 161.
5. Ibid., p. 161.
6. Ibid., p. 168.
7. Stephen Fox, *John Muir and His Legacy* (Boston: Little, Brown, 1981), p. 121.
8. Ibid., pp. 145–46.
9. Lester W. Milbrath, *Environmentalists: Vanguard for a New Society* (Albany: State University of New York Press, 1984), p. 72.
10. Bryan G. Norton, "Conservation and Preservation: A Conceptual Rehabilitation," *Environmental Ethics* 8 (1986): 195–220.
11. Arne Naess, "The Shallow and the Deep, Long-Range Ecology Movement. A Summary," *Inquiry* 16 (1973): 95–100.
12. See Peter Borelli, "Environmentalism at a Crossroads," *Amicus,* Summer (1987): 34.
13. One might call this general approach "philosophical pragmatism," although I hope that the analysis presented here will be of interest to readers who reject pragmatism as a philosophical theory and also to those who care little about philosophical theory at all. See Preface.

257

Chapter 2

1. Gifford Pinchot, *Breaking New Ground* (Washington, D.C.: Island Press, 1987; originally published 1947), p. 100.
2. Roderick Nash, *Wilderness and the American Mind*, 3rd ed. (New Haven: Yale University Press, 1982), p. 137.
3. Pinchot, *Breaking New Ground*, p. 103.
4. Holway R. Jones, *John Muir and the Sierra Club* (San Francisco: Sierra Club Books, 1965), p. 18; Nash, *Wilderness*, pp. 134–38; Lawrence Rakestraw, "Sheep Grazing in the Cascade Range: John Minto *vs.* John Muir," *Pacific Historical Review* 27 (1958): 371–82.
5. Stephen Fox, *John Muir and His Legacy* (Boston: Little, Brown, 1981), p. 33. Also see Tom Melham, *John Muir's Wild America* (Washington, D.C.: National Geographic Society, 1976), p. 36.
6. Fox, *John Muir*, pp. 39–40.
7. Ibid., p. 43.
8. Ibid.
9. Ibid.; quoted from an account published in the *Boston Recorder*, December 21, 1866.
10. Ibid., p. 44. Again, quoted from the *Boston Recorder*.
11. John Muir, *A Thousand-Mile Walk to the Gulf* (Boston: Houghton Mifflin, 1981; originally published 1916), p. 70.
12. Fox, *John Muir*, p. 52, quoted from the John Muir Papers, Journal, Fall, 1867, pp. 150–56.
13. Fox, *John Muir*, p. 53.
14. Quoted in Melham, *John Muir's America*, p. 20.
15. Fox, *John Muir*, pp. 70–80; also see Roderick Nash, *The Rights of Nature* (Madison: The University of Wisconsin Press, 1989), pp. 40–41.
16. Muir's rejection of Christian anthropocentrism has led, recently, to a lively scholarly disagreement, pitting scholars who interpret Muir as having rejected Christianity (these include Stephen Fox and Roderick Nash) against those who see Muir as proposing an interesting reinterpretation of Christianity. (See J. Baird Callicott, "Genesis Revisited: Muirian Musings on the Lynn White, Jr. Debate," *Environmental History Review* 14 (1990): 65–90 for a discussion of Muir's "reinterpretation.") A disadvantage of the former interpretation, which is especially evident in Nash's treatment of Muir, is that Muir, throughout his published writings, continued to use religious and Christian-sounding language. Nash, therefore, is driven to accuse Muir of deliberate deceit and a failure to give his real reasons for preservationist sentiments in his many published writings.
17. Fox, *John Muir*, p. 7.
18. Ibid., p. 49.
19. Quoted ibid., p. 107.
20. Pinchot, *Breaking New Ground*, pp. 2–3.
21. Ibid., p. 3.
22. M. Nelson McGeary, *Gifford Pinchot, Forester-Politician* (Princeton: Princeton University Press, 1960), p. 19; Pinchot, *Breaking New Ground*, pp. 7–22.
23. Pinchot, *Breaking New Ground*, p. 11.
24. Ibid., p. 27.

25. Ibid.
26. Ibid., p. 322.
27. Ibid.
28. Ibid., p. 323.
29. Ibid., pp. 325–26.
30. Ibid., p. 326.
31. Ibid.
32. McGeary, *Gifford Pinchot*, p. 86.
33. Fox, *John Muir*, p. 121.
34. Ibid., p. 109.
35. Jones, *John Muir*, p. 4.
36. Fox, *John Muir*, pp. 116–17. Also see Peter J. Schmidt, *Back to Nature* (New York: Oxford University Press, 1969).
37. Pinchot, *Breaking New Ground*, p. 28.
38. Ibid., pp. 28–29.
39. Ibid., p. 27.
40. Ibid., p. 28.
41. Ibid., pp. 26–27.
42. Ibid.
43. Ibid., p. 29.
44. McGeary, *Gifford Pinchot*, p. 27.
45. Ibid., p. 39.
46. Ibid., p. 41.
47. Pinchot, *Breaking New Ground*, pp. 107–8.
48. Fox, *John Muir*, p. 112.
49. Pinchot, *Breaking New Ground*, p. 109.
50. McGeary, *Gifford Pinchot*, p. 40.
51. Fox, *John Muir*, p. 113.
52. Pinchot, *Breaking New Ground*, p. 122.
53. McGeary, *Gifford Pinchot*, p. 54.
54. Quoted ibid., p. 61.
55. Ibid., pp. 54–55.
56. Ibid., p. 42.
57. Pinchot, *Breaking New Ground*, p. 177.
58. Ibid., p. 179.
59. Ibid., p. 180.
60. Ibid., p. 181.
61. Fox, *John Muir*, p. 144.
62. Ibid., p. 144.
63. Quoted ibid., p. 115.
64. Pinchot, *Breaking New Ground*, pp. 102–3.
65. Fox, *John Muir*, p. 81.
66. Eugene C. Hargrove, *Foundations of Environmental Ethics* (Englewood Cliffs, NJ: Prentice-Hall, 1989), pp. 84–88.
67. Ibid., pp. 86–87. See also Allen Carlson, "Nature and Positive Aesthetics," *Environmental Ethics* 6 (Spring 1984).
68. Ibid., p. 88.
69. Ibid., p. 91.
70. Fox, *John Muir*, p. 81.

71. Muir, *Thousand-Mile Walk*, p. 98.
72. John Muir, "Wild Wool," in *Wilderness Essays* (Salt Lake City: Peregrine Smith, 1980), p. 236.
73. Fox, *John Muir*, p. 82.
74. Ibid., p. 23. Quoted from John Muir, *Overland Monthly*, August 1873.
75. Nash, *Wilderness*, p. 136.
76. Quoted ibid., pp. 136–37.

Chapter 3

1. Aldo Leopold, *A Sand County Almanac* (London: Oxford University Press, 1949), p. viii.
2. See Curt Meine, *Aldo Leopold: His Life and Work* (Madison: University of Wisconsin Press, 1988), p. 188.
3. Aldo Leopold, "Some Fundamentals of Conservation in the Southwest," *Environmental Ethics* 1 (Summer 1979): 140.
4. Ibid., p. 139.
5. Ibid.
6. Ibid., p. 141.
7. Such as W. T. Hornaday, who waged war against "game hogs" from his post as director of the Bronx Zoo. Leopold, who supported hunting, and Hornaday, who believed it must be curtailed, united to support game preserves for the perpetuation of wildlife populations. See Stephen Fox, *John Muir and His Legacy* (Boston: Little, Brown, 1981) and Meine, *Aldo Leopold*, p. 262.
8. Leopold, "Some Fundamentals," p. 141.
9. Arthur Twining Hadley, *Some Influences in Modern Philosophic Thought* (New Haven, CT: Yale University Press, 1913), p. 71. A. T. Hadley studied political economy at the University of Berlin and returned to become a tutor and later a professor at Yale, where he had been an undergraduate. Noted for the breadth of his knowledge, he offered classes on economics and political ethics that were extremely popular. Hadley became the first lay president of Yale in 1899. He described himself as a "thorough-going pragmatist" and generally quoted William James's work as representative of modern philosophical thinking. (Morris Hadley, *Arthur Twining Hadley* [New Haven, CT: Yale University Press, 1948], p. 197.)
10. A. T. Hadley, *Some Influences*, pp. 121–26.
11. Leopold, "Some Fundamentals," pp. 139–140.
12. Ibid., pp. 140–411. Leopold used the term "anthropomorphic" as we today use "anthropocentric"—to refer to a value system that bases all value in human motives. Except in quotations of Leopold, I will follow current practice and use "anthropocentrism."
13. Ibid., p. 136.
14. Fortunately, there exist two excellent, detailed treatments of Leopold's changing approaches to deer and wolf management. See Susan Flader, *Thinking Like A Mountain* (Lincoln: University of Nebraska Press, 1974) and Meine, *Aldo Leopold*, p. 181. Quotations are taken from Meine, who quotes Leopold's 1920 speech "The Game Situation in the Southwest."
15. Meine, *Aldo Leopold*, p. 453.
16. Aldo Leopold, *Game Management* (Madison: University of Wisconsin Press, 1986; first published in 1933), p. 392.

17. Meine, *Aldo Leopold,* p. 244.
18. Leopold, "Some Fundamentals," pp. 136–37.
19. Ibid., p. 137.
20. I have discussed these concepts in detail in Bryan G. Norton, *Why Preserve Natural Variety?* (Princeton, NJ: Princeton University Press, 1986), Chap. 5.
21. Quoted in Meine, *Aldo Leopold,* pp. 353–54, from "Deer and Dauerwald in Germany, Part I: History," *Journal of Forestry* (1936): 374.
22. Quoted in Meine, *Aldo Leopold,* p. 368.
23. Ibid., p. 369. Quoted from "Conservation in Mexico," *American Forests* 43 (1937): 120.
24. Meine, *Aldo Leopold,* p. 453.
25. Leopold, *Sand County,* pp. 129–33.
26. See Meine, *Aldo Leopold,* for an account of Leopold's fledgling attempts at studying the game situation empirically, p. 458, and of Leopold's trials and tribulations in the politics of deer management, pp. 442f.
27. Ibid., p. 458. Quoted from "The 1944 Game Situation," *Aldo Leopold Papers,* March 24, 1944.
28. Leopold, *Sand County,* p. 132.
29. Meine, *Aldo Leopold,* pp. 282–84.
30. Leopold, *Sand County,* p. 215.
31. Aldo Leopold, "A Biotic View of Land," *Journal of Forestry* 37 (1939): 727–28.
32. Talbot Page, *Conservation and Economic Efficiency* (Baltimore: Johns Hopkins University Press, 1977). Leopold, like Page, insisted that the conservation criterion is primary in that policies unacceptable for conservation reasons are eliminated from further consideration. Page and Leopold also agree that the system of analysis cannot be simply aggregative. The operation of economic analysis is limited to particular, productive subsystems. Analyses of productivity data within cells is not aggregated with data concerning the health of the larger context. Unlike Page, who applies this reasoning only to a "materials policy," Leopold applied similar reasoning to all management problems. Leopold also argued that the conservation criterion should be ecologically informed and conceived on an organic model, including a strongly normative conception of "ecological health," capable of determining which activities contribute to illness in the environmental context of the activities undertaken.
33. Leopold, "A Biotic View," p. 727. This essay, an extremely important one, represents a summary of what Leopold had learned about environmental management and explicitly refers to all three of the "paradigms of failed management" that I have discussed here.
34. Ibid.
35. Ibid., p. 729.
36. Cf. George Woodwell, "On the Limits of Nature," in Robert Repetto, ed., *The Global Possible* (New Haven: Yale University Press, 1985).
37. Leopold, "A Biotic View," p. 728.
38. Leopold, *Sand County,* p. 221.
39. Ibid., pp. viii–ix.
40. Ibid., p. 167.
41. Ibid., p. 176.
42. Ibid., p. 173.
43. Ibid., p. 177.
44. Ibid., pp. 177–78.

45. Ibid., p. 178.
46. Ibid., p. 179.
47. Ibid., p. 181.
48. Ibid., pp. 200–201.
49. See, for example, John Passmore, *Man's Responsibility for Nature* (New York: Charles Scribner's Sons, 1974), Chap. 5 and Joseph Petulla, *American Environmentalism: Values, Tactics, and Priorities* (College Station: Texas A&M University Press, 1980), pp. 16, 204–5. More recently, Callicott has interpreted the land ethic as based on "ecocentric values," values "inherent" in the land community, but not precisely equal to the value placed on humans. See J. Baird Callicott, "The Metaphysical Implications of the Land Ethic," in *In Defense of the Land Ethic* (Albany: State University of New York Press, 1989), pp. 101–14.
50. Leopold, *Sand County*, p. 221.

Chapter 4

1. Curt Meine, *Aldo Leopold: His Life and Work* (Madison: University of Wisconsin Press, 1988), p. 523.
2. Stephen Raushenbush, "Conservation in 1952," *The Annals of the American Academy of Political and Social Science* 281 (1952): 1.
3. Grant McConnell, "The Conservation Movement—Past and Present," *Western Political Quarterly* 7 (1954): 463.
4. Samuel Hays, *Beauty, Health, and Permanence: Environmental Politics in the United States, 1955–1985* (Cambridge: Cambridge University Press, 1987), pp. 3–4, 22–26.
5. Robert Mitchell, "From Conservation to Environmental Movement: The Development of the Modern Environmental Lobbies," Resources for the Future Discussion Paper QE85-12, June 1985, p. 3.
6. Ibid., pp. 6–7.
7. Donald Fleming, "Roots of the New Conservation Movement," *Perspectives in American History* 6 (1972): 28.
8. Ibid., p. 29.
9. See, for example, Barry Commoner, *The Closing Circle* (New York: Knopf, 1971); Paul R. Ehrlich, *The Population Bomb* (New York: Ballantine Books, 1968); Donella H. Meadows et al., *The Limits of Growth* (New York: Universe Books, 1972).
10. Mitchell, "From Conservation," pp. 1–29.
11. Council on Environmental Quality et al., *Public Opinion on Environmental Issues* (Washington, D.C.: U.S. Government Printing Office, 1980).
12. Council on Environmental Quality et al., *Public Opinion*, p. 1.
13. Ibid., p. 5.
14. Ibid., pp. 6–8. They chose among ten national problems including also reducing the amount of crime, conquering "killer" diseases, and so on.
15. Ibid., p. 8.
16. Andrew Kohut and James Shriver, "The Environment: Environment Regaining a Foothold on the National Agenda," *Gallup Report*, June 1989, pp. 2–3, 6.
17. Robert Cameron Mitchell, "Public Opinion and Environmental Politics in the 1970s and 1980s," in Norman J. Vig and Michael E. Kraft (eds.), *Environmen-*

tal Policy in the 1980s (Washington, D.C.: Congressional Quarterly Press, 1984), pp. 60–67 passim.

18. Riley E. Dunlap, "Trends in Public Opinion Toward Environmental Issues: 1965–1990," presented at the annual meeting of the American Association for the Advancement of Science, New Orleans, February 15–20, 1990, Table 4.

19. Dennis A. Gilbert, *Compendium of American Public Opinion* (New York: Facts on File Publications, 1988), p. 121.

20. Ibid., pp. 124–25.

21. Stephen Fox, *John Muir and His Legacy* (Boston: Little, Brown, 1981), p. 333.

22. John Passmore, *Man's Responsibility for Nature* (New York: Charles Scribner's Sons, 1974), p. 73.

23. Ibid.

24. Ibid.

25. Ibid., p. 101.

26. Arne Naess, "The Shallow and the Deep, Long-Range Ecology Movement. A Summary," *Inquiry* 16 (1973).

27. Ibid., p. 100.

28. Warwick Fox, "Approaching Deep Ecology: A Response to Richard Sylvan's Critique of Deep Ecology," *Environmental Studies Occasional Paper #20*, University of Tasmania, Australia, 1986, pp. 7–9.

29. It should be noted that, while many classifications of environmentalists are dualistic, there also appear tripartite classifications. George R. Hall divides environmentalists ("conservationists" in his terminology) into (1) neo-Malthusians, (2) technoconservationists, and (3) naturalist-preservationists; Joseph Petulla recognizes (1) a biocentric perspective, (2) an ecologic perspective, and (3) an economic perspective. While Hall's (2) and Petulla's (3) seem closely related, the remaining categories do not appear commensurate. George R. Hall, "Conservation as a Public Policy Goal," in Dennis Thompson (ed.), *Politics, Policy, and Natural Resources* (New York: Free Press, 1972), pp. 181–91; Joseph M. Petulla, *American Environmentalism: Values, Tactics, and Priorities* (College Station: Texas A&M University Press, 1980), pp. 25–39.

30. The idea of a DSP was introduced by D. C. Pirages and P. R. Ehrlich in *Ark II: Social Response to Environmental Imperatives* (San Francisco: W. H. Freeman, 1974). This concept was contrasted with the NEP, operationalized and tested by R. E. Dunlap and K. D. Van Liere in "The 'New Environmental Paradigm,'" *Journal of Environmental Education* 9 (1978): 10–19. It is important to distinguish between the *descriptive usefulness* of these categorizations to separate environmentalists in opposed camps from the *philosophical usefulness* of a distinction between anthropocentric and biocentric value systems. The latter, philosophical distinction suggests an empirical search for two exclusive groups of environmentalists only if the two value systems are defined in exclusive terms (such that it would be inconsistent to embrace both types of value). We shall return to the philosophical usefulness of an exclusive dichotomy in Chapter 12.

31. Bill Devall, "The Deep Ecology Movement," *Natural Resources Journal* 20 (April 1980): 300.

32. Ibid., p. 302.

33. William Devall and George Sessions, *Deep Ecology: Living As if Nature Mattered* (Salt Lake City: Peregrine Smith Books, 1985).

34. See Stephen Toulmin, *Foresight and Understanding* (New York: Harper & Row,

1961) and Thomas Kuhn, *The Structure of Scientific Revolutions* (Chicago: University of Chicago Press, 1962).

35. Kuhn has denied that he intended this feature, at least in any extreme form. See "Postscript—1969" in the second edition of *Structure*. Without adjudicating claims regarding Kuhn's intent, it seems fair to say that his book would have aroused less excitement and debate had he not been so interpreted.

36. Riley E. Dunlap and Kent D. Van Liere, "Commitment to the Dominant Social Paradigm and Concern for Environmental Quality," *Social Science Quarterly* 65 (December 1984: 1013, quoting Dennis C. Pirages, "Introduction: A Social Design for Sustainable Growth," in Dennis C. Pirages (ed.), *The Sustainable Society* (New York: Praeger, 1977)—my emphasis.

37. Dunlap and Van Liere, "Commitment," p. 1015. Dunlap and Van Liere operationalize agreement with a dominant social paradigm according to a constellation of eight factors, which are heavily weighted with policy commitments. Their characterization contains no direct reference to value issues. It is not clear, therefore, that their data bears directly on more value-based differences characterized in Devall's categories.

38. Ibid.

39. Lester W. Milbrath, *Environmentalists: Vanguard for a New Society* (Albany: State University of New York Press, 1984). Milbrath provides detailed comparative data on the United States, Great Britain, and West Germany. His data, gathered in a massive, three-nation research effort, provides the most comprehensive information on attitudes toward the environment and environmental issues. My criticisms, which are developed in response to Milbrath's interpretation of one subset of his data (1980 U.S. data), can, I think, be generalized to apply to Milbrath's entire interpretative structure. It should be noted, however, that my criticisms do not, for the most part, call into question the validity of Milbrath's data, but only his interpretation of it.

40. See Milbrath, *Environmentalists*, p. 22, for the complete list of components. It is worth noting, in passing, that the three subcomponents of Milbrath's value category are logically (and, I would argue, psychologically) independent. One could (B) wish to live harmoniously with nature and (C) prefer environmental protection to economic growth for human-oriented reasons (one might believe, for example, that these policies would lead to more fulfilling human lives). These subcomponents are therefore independent of (A) and supporters of (B) and (C) should not be interpreted as necessarily supporting (A).

41. Members of the Sierra Club, the Nature Conservancy, Friends of the Earth, and the Environmental Defense Fund. He then supplemented this list with names from the *Conservation Directory*, choosing the latter individuals deliberately to overrepresent "nature conservationist type organizations such as hunting and fishing clubs."

42. Milbrath, *Environmentalists*, p. 26.

43. Ibid., p. 25.

44. Ibid., p. 14.

45. Ibid., p. 26 (emphasis mine).

46. See ibid., p. 28. Also see p. 44 (where Milbrath explains that he obtained a "neat linear progression from left to right across the [dimension measuring postures toward social change]." It is unclear how he mapped this "neat progression" into subgroups of environmentalists that he posits. Also see note 3, p. 64,

where he admits a certain arbitrariness in the choice of variables used for this purpose.

47. In Milbrath's defense it should be mentioned that he seems almost to recognize this, cautioning that "the real world, of course, is not so neatly structured and divided [as a paradigmatic analysis would suggest]. Most studies have shown that many people hold beliefs that partake of both paradigms; the reader should keep this qualifier in mind throughout the following discussion" (p. 21). But one might ask, considering the way in which Milbrath reifies subcategories of environmentalists according to their acceptance or rejection of the DSP, whether Milbrath has heeded his own warning.

48. See Bryan G. Norton, "Conservation and Preservation: A Conceptual Rehabilitation," *Environmental Ethics* 8 (1986): 200–202, for a more detailed discussion of the proposed definitions.

Chapter 5

1. Twentieth-century Western philosophy, both Anglo-American and Continental, has recognized the powerful effect that language and conceptualization have upon thought and action. While many who would call themselves realists still assert the dominant role of real objects in concept formation, few any longer doubt the significant impact of theoretical and value precepts on thought, or the close connection of all of these with action. See, for example, Ludwig Wittgenstein, *Philosophical Investigations* (Garden City, NY: Anchor Books, 1966). For a more accessible account of worldviews, see Thomas W. Overholt and J. Baird Callicott, *Clothed in Fur and Other Tales: An Introduction to an Ojibwa World View* (Washington, DC: University Press of America, 1982).

2. R. G. Collingwood, *An Essay on Metaphysics* (Oxford: Oxford University Press, 1940). Collingwood argues, in this classic treatment, that persons and cultures are guided by sets of "ultimate presuppositions." He does not, incidentally, use the more recent terms "worldview" or "paradigm" to characterize sets of such presuppositions.

3. See William Burch, *Daydreams and Nightmares* (New York: Harper & Row, 1971), pp. 35–36, for a discussion of the attitudes of the early colonists as typical of cultures facing newly opened frontiers.

4. See Daniel Boorstin, *The Americans: The Colonial Experience* (New York: Random House, 1958), p. 250, for a discussion of five reasons explaining this carelessness.

5. See John Locke, *The Second Treatise of Government*, chap. 5, for a discussion of how unowned natural objects—he uses the example of acorns—gain value through human labor.

6. Roderick Nash, *Wilderness and the American Mind*, 3rd ed. (New Haven: Yale University Press, 1982), p. 31.

7. Perry Miller, *An Errand into the Wilderness* (Cambridge: Harvard University Press, 1956).

8. David E. Shi, *The Simple Life* (Oxford: Oxford University Press, 1985), p. 11.

9. This conclusion is explained by growth-oriented economists today not by claims of unlimited abundance, but by a faith in the unlimited substitutability of one resource for another. See Bryan G. Norton, "Sustainability, Human Welfare,

and Ecosystem Health," in preparation, for a fuller discussion of the role of this idea in determining views on resource use.

10. I am referring here to Carson's writing career and contribution as an environmentalist, not to her career as a biological scientist.

11. For a more detailed discussion of these epistemological points, see Stephen Toulmin, *Human Understanding*, Vol. I (Princeton: Princeton University Press, 1972), pp. 41–144, 478–503.

12. As in the case of one particular version of the diversity-stability hypothesis. See Bryan G. Norton, *Why Preserve Natural Variety?* (Princeton: Princeton University Press, 1987).

13. Robert Mitchell, "Since Silent Spring: The Institutionalization of Counter-Expertise by the United States Environmental Law Groups" (Washington, D.C.: Resources for the Future, 1979).

14. The situation was first described to me by Spencer Carr.

15. The idea of a self-sufficient philosophy is closely tied to the idea of foundationalism in epistemology. See Chapter 12 for a discussion of these issues in connection with the ideology of deep ecology.

16. John Rawls, *A Theory of Justice* (Cambridge: Harvard University Press, 1971), pp. 19ff.

17. Ibid., p. 19.

18. Ibid., pp. 20–21.

19. Rawls notes that in presenting his theory in book form, he presents only the conclusions of such a process—the principles as already worked out (p. 21). Here, we are doing practical philosophy and must include at least some of the trips around the circle in order to understand the true interanimation of values and policy positions.

20. "Applied philosophy" is therefore a misnomer. One can do "pure" philosophy—as in the fortress example—or one can do "practical" philosophy—as we are doing in this book. But one cannot finish one's pure philosophy and then apply it to the real world without adjustment.

21. See John Passmore, *Man's Responsibility for Nature* (New York: Charles Scribner's Sons, 1974), pp. 115–18.

22. See Christopher D. Stone, *Earth and Other Ethics* (New York: Harper & Row, 1987).

23. The reader may, at this point, cynically, but not without justification, remind me that, in the last section, I advised against the use of hypothetical examples. Chagrined, I can only reply that my use of a hypothetical case is not, here, designed to draw conclusions about values, but only to introduce a model. The true value of the model will then be tested, in subsequent chapters, against real cases—and it will be judged a worthwhile model only if it illuminates those real cases.

24. It should be noted, however, that this first stage can also be, and increasingly is, much more proactive—as when an opportunity to restore damaged ecological systems leads to public support for restoration activities.

Chapter 6

1. William Tucker, *Progress and Privilege* (Garden City, NY: Anchor Press/Doubleday, 1982). The elitism charge has also been a popular theme of certain industry groups, especially when they have attempted to enlist the support of

labor and community-development organizations. See Samuel P. Hays, *Beauty, Health, and Permanence: Environmental Politics in the United States, 1955–1985* (Cambridge: Cambridge University Press, 1987), chap. 9.

2. See Robert Cameron Mitchell, "How 'Soft,' 'Deep,' or 'Left?' Present Constituencies in the Environmental Movement for Certain World Views," *Natural Resources Journal* 20 (April 1980): 345–58.

3. See, for example, Richard Levins and Richard Lewontin, *The Dialectical Biologist* (Cambridge: Harvard University Press, 1985). They say that it is "ironic" that environmentalists oppose bourgeois society, but in doing so, emphasize the same "stability and complexity of natural communities of species to oppose the expansion of the very capitalist system of production that gave rise to the ideology originally" (p. 22).

4. Rockefeller Brothers Fund, *The Unfinished Agenda: The Citizen's Guide to Environmental Issues* (New York: Crowell, 1977), pp. 156–57.

5. Robert Cahn (ed.), *An Environmental Agenda for the Future* (Washington, DC: Island Press, 1985), p. 7.

6. See Herman Kahn, *The Coming Boom: Economic, Political, and Social* (New York: Simon and Schuster, 1982); Julian L. Simon, *The Ultimate Resource* (Princeton: Princeton University Press, 1981); Julian L. Simon and Herman Kahn (eds.), *The Resourceful Earth: A Response to Global 2000* (Oxford: Blackwell, 1984).

7. See T. H. Watkins, "Untrammeled by Man," *Audubon*, November 1989, pp. 78ff., for a history of this episode in Forest Service history.

8. Interview with author, Washington, D.C., May 21, 1986; also see Hays, *Beauty, Health*.

9. William K. Wyant, *Westward in Eden: The Public Lands and the Conservation Movement* (Berkeley: University of California Press, 1982), pp. 279–81.

10. See Hays, *Beauty, Health*, chap. 4, especially pp. 123–28, for a more detailed account of how the Forest Service, although showing bureaucratic resistance, has gradually accepted that its task is one of "a broker among conflicting demands" (p. 128).

11. See Alston Chase, *Playing God in Yellowstone* (Boston: Atlantic Monthly Press, 1986), for a detailed account of the problems caused when the Park Service designed management plans on the (false) assumption that Yellowstone National Park could be managed as a self-sufficient ecosystem.

12. See Amory Lovins, *Soft Energy Paths: Toward a Durable Peace* (San Francisco: Friends of the Earth International 1977 [distributed by Ballinger]).

13. Kristen S. Shrader-Frechette, "Ethics and Energy," in Tom Regan (ed.), *Earthbound* (New York: Random House, 1984), pp. 109–10.

14. The 1977 report of the Workshop on Alternative Energy Strategies (WAES), the 1977 and 1983 versions of the World Energy Conference (WEC) report on world energy demand, and the 1981 Report of the Energy Systems Program Group of the International Institute for Applied Systems Analysis (IIASA). See Amulya N. Reddy, "Energy Issues and Opportunities," in Robert Repetto (ed.), *The Global Possible* (New Haven: Yale University Press, 1985), pp. 371–73.

15. Ibid., p. 372.

16. Lovins, *Soft Energy Paths*, p. 6.

17. See Ibid., p. 56; Barbara Ward, *Progress for a Small Planet*, (New York: Norton, 1982), p. 31; and Shrader-Frechette, "Ethics and Energy," p. 110.

18. Lovins, *Soft Energy Paths*, p. 46.

19. Cahn, *An Environmental Agenda for the Future*, pp. 41–42.
20. Mitchell, "How 'Soft,' " p. 347.
21. Lovins, *Soft Energy Paths*, pp. 28–29.
22. Ibid., p. 9.
23. Ibid., p. 23.
24. Ibid., pp. 4, 8.
25. Also see Shrader-Frechette, "Ethics and Energy," p. 111.
26. Ibid.
27. Ibid., p. 112.
28. Ibid., p. 115.
29. George Woodwell, "On the Limits of Nature," in Repetto (ed.), *The Global Possible*, p. 48.
30. Cahn, *An Environmental Agenda for the Future*, p. 38.
31. The latter approach sometimes appears in an extreme form, such as the "life-boat ethic," whose supporters advocate abandoning some nations as "lost causes" and emphasize closing borders and even cutting off economic aid to nations with high birth rates. See Garrett Hardin, "Living on a Lifeboat," *BioScience* 24 (October 1974). But mainline environmentalists have generally rejected this extreme position.
32. Bryan G. Norton, *Why Preserve Natural Variety?* (Princeton: Princeton University Press, 1987), chap. 1, 10, and 11.
33. Bryan G. Norton, "Thoreau's Insect Analogies: Or, Why Environmentalists Hate Mainstream Economists," *Environmental Ethics* 13/3 (1991).
34. Reddy, "Energy Issues," p 373; also see the World Commission on Environment and Development, *Our Common Future* (Oxford: Oxford University Press, 1987), chap. 2.
35. Interview with author, July 28, 1987.
36. Lovins, *Soft Energy Paths*, p. 9.
37. Rockefeller, Brothers Fund, *Unfinished Agenda*, pp. 156–57.
38. See Chapter 3.
39. For an excellent explanation, in economists' terminology, of a similar, two-step system, see Talbot Page, *Conservation and Economic Efficiency* (Baltimore: Johns Hopkins University Press, 1977). Also, refer back to Chapter 3 of this book.

Chapter 7

1. Robert Mitchell, Unpublished data from mail survey of members of five national environmental groups, 1978.
2. U.S. Department of Commerce, *Statistical Supplement to the Survey of Current Business* (Washington, D.C.: U.S. Government Printing Office, 1963), pp. 26, 28.
3. Rachel Carson, *Silent Spring* (New York: Fawcett Publications, 1962), pp. 13–14.
4. Frank Graham, *Since Silent Spring* (Boston: Houghton-Mifflin, 1970), p. 75.
5. Ibid., pp. 69–75.
6. In particular, Carson almost surely overstated the case for bioaccumulation as pesticides travel up the food chain. See Graham, above, for a detailed evaluation of the scientific debate.
7. Carson, *Silent Spring*, p. 22.

8. Ibid., p. 22.

9. Richard Ayers, interview with author, Washington, D.C., August 4, 1986.

10. Milton Russell, "Technology, Science, Risk and Environmental Protection," Address delivered at Oak Ridge National Laboratory, November 20, 1985; also see Paul Portney, *Current Issues in Natural Resource Policy* (Washington, D.C.: Resources for the Future, 1982), chap. 11.

11. Carolyn Lockhead, "Pollutants Reined by Market Rules," *Insight*, July 3, 1989, p. 9.

12. See Alan Weisman "L.A. Fights for Breath," *New York Times Magazine*, July 30, 1989.

13. Clean Air Act of 1970, Public Law 91-604, December 31, 1970.

14. Federal Water Pollution Control Act Amendments of 1972 and 1977, Public Law 92-500, October 18, 1972, and Public Law 95-217, December 27, 1977.

15. Russell, "Technology," p. 4.

16. See Carolyn Lockhead, "Credit Bartering in the Market for Air Pollution," *Insight*, July 3, 1989, pp. 15–17, for a more detailed account of the various plans proposed.

17. William Butler, interview with author, April 4, 1986.

18. Ibid.

19. Terry Davies, former Assistant Administrator for Policy and Planning at EPA, interview with author, July 31, 1986.

20. Paul Portney, *Public Policies for Environmental Protection* (Washington, D.C.: Resources for the Future, 1990), chap. 1.

21. Ayers, interview with author.

22. Ibid.

23. Ibid.

24. Lockhead, "Pollutants Reined by Market Rules," p. 13. In a personal communication, Portney clarified this quotation as not implying that standard-setting based on noneconomic factors is preferable.

25. Paul Portney, interview with author, Washington, D.C., August 6, 1986.

26. Ayers, interview with author.

27. Portney, interview with author.

28. Robert Cahn (ed.), *An Environmental Agenda for the Future* (Washington, DC: Island Press, 1985), p. 69.

29. Davies, interview with author.

Chapter 8

1. The presentations of the participants are included in E. O. Wilson (ed.), *Biodiversity* (Washington, D.C.: National Academy Press, 1988), Part 5.

2. For a fuller discussion of this case, see Philip Shabecoff, "New Battles Over Endangered Species," *New York Times Magazine*, June 4, 1978; also see Bryan G. Norton, *Why Preserve Natural Variety?* (Princeton: Princeton University Press, 1987), pp. ix–x.

3. For a dramatic account of these developments see Christopher Stone, *Should Trees Have Standing?* (Los Altos, CA: William Kaufmann, 1974).

4. See, for example, Laurence H. Tribe, "Ways Not to Think About Plastic Trees: New Foundations for Environmental Law," *The Yale Law Journal* 83 (June 1974): 1315–48; and Mark Sagoff, "On the Preservation of Species," *Columbia Journal of Environmental Law* 7 (Fall 1980): 33–67.

5. A. C. Fisher and W. M. Hanemann, "Option Value and the Extinction of Species," California Agricultural Experiment Station, Berkeley. Also see Bryan Norton, "Commodity, Amenity, and Morality: The Limits of Quantification in Valuing Biodiversity," in Wilson (ed.), *Biodiversity*, p. 202.

6. For a detailed discussion of further disagreements between Aggregators and Moralists regarding details of value assignments, see Norton, "Commodity, Amenity, and Morality" and *Why Preserve Natural Variety?*

7. Anne H. and Paul R. Ehrlich, "Population and Development Misunderstood," *The Amicus Journal* 8 (Summer 1986): 8–10.

8. See David Ehrenfeld, "The Conservationist's Dilemma," in *The Arrogance of Humanism* (New York: Oxford University Press, 1978), also see Aldo Leopold, *A Sand County Almanac*, (London: Oxford University Press, 1949).

9. Thomas Lovejoy, "Species Leave the Ark One by One," in Bryan G. Norton (ed.), *The Preservation of Species* (Princeton: Princeton University Press, 1986), p. 22.

10. In congressional testimony, Peter Raven has said, "It stands to reason that the extinction of a single kind of plant has the potential of bringing about the extinction of 12 to 15 or more kinds of animals and other dependent organisms," in Hearings before the Subcommittee on Fisheries and Wildlife Conservation and the Environment of the House of Representatives' Committee on Merchant Marine and Fisheries, February 22, March 8, 1982. Serial No. 97-32, U.S. Government Printing Office, Washington, D.C., 1982, p. 124.

11. See Scott Nixon, "Between Coastal Marshes and Coastal Waters—A Review of Twenty Years of Speculation and Research on the Role of Salt Marshes in Estuarine Productivity and Water Chemistry," in Peter Hamilton and Keith B. MacDonald (eds.), *Estuarine and Wetland Processes with Emphasis on Modelling* (New York: Plenum Press, 1980), pp. 437–525.

12. I have developed this argument in detail in Bryan G. Norton, "Thoreau's Insect Analogies: Or, Why Environmentalists Hate Mainstream Economists," *Environmental Ethics* 13/3 (1991).

13. Leopold, "A Biotic View of Land," *Journal of Forestry*, 37 (1939): 728.

14. Robert H. Whittaker, *Communities and Ecosystems* (New York: Macmillan, 1970), p. 103.

15. Whittaker used the less intuitive terms "alpha," "beta," and "gamma diversity" as designators of the levels. Robert H. Whittaker, "Vegetation of the Siskiyou Mountains, Oregon and California," *Ecological Monographs* 30 (1960): 30ff. Also see Robert MacArthur, "Patterns of Species Diversity," *Biological Review* 40 (1965): 510–33.

16. Terry L. Erwin, "The Tropical Forest Canopy: The Heart Of Biotic Diversity," in Wilson (ed.), *Biodiversity*, p. 128.

17. Leopold, *Sand County*, pp. 224–25.

18. Ibid., p. 224. See T. F. H. Allen and Thomas B. Starr, *Hierarchy: Perspectives for Ecological Complexity* (Chicago: University of Chicago Press, 1982), pp. 8–10.

19. Arthur Koestler, *The Ghost in the Machine* (New York: Macmillan, 1967). p. 343.

20. Leopold, *Sand County* p. 133; also see p. 206.

21. Allen and Starr, *Hierarchy*, pp. 11–13.

22. Ibid., p. 15.

23. H. H. Pattee, *Hierarchy Theory: The Challenge of Complex Systems* (New York: Braziller, 1973), pp. 105–7.

24. Allen and Starr, *Hierarchy*, p. 219.

25. Ibid., p. 10.

26. R. V. O'Neill, et al., *A Hierarchical Concept of Ecosystems* (Princeton: Princeton University Press, 1986), p. 39.

27. See Tom Regan, "The Nature and Possibility of an Environmental Ethic," *Environmental Ethics* 3 (Spring 1981). Regan argues that it is "an improper question" to compare the disvalue of the loss of caribou lives with a given amount of economic dislocation. He concludes, "because these two kinds of good are incommensurable," an ethic that "endeavors to accommodate both kinds of goodness is doomed to fail," p. 32.

28. S. V. Ciriacy-Wantrup, *Resource Conservation: Economics and Politics* (Berkeley and Los Angeles: University of California Division of Agricultural Sciences, 1959).

29. See Laura M. Lake, *Environmental Regulation: The Political Effects of Implementation* (New York: Praeger, 1982), pp. 89–92, for a brief history of specific court decisions that developed this liberalizing trend regarding access to the courts.

Chapter 9

1. Curt Meine, *Aldo Leopold: His Life and Work* (Madison: University of Wisconsin Press, 1988), p. 454.

2. Thomas M. McNamee, "Putting Nature First: A Proposal for Whole Ecosystem Management," *Orion Nature Quarterly* 5 (1986): 6.

3. Ibid., p. 8.

4. Ibid., p 7.

5. Alston Chase, *Playing God in Yellowstone* (Boston: Atlantic Monthly Press, 1986), pp. 28–30.

6. Ibid., pp. 11–13, 28. The research on the causes of beaver decline was done by Robert Jonas.

7. Ibid., pp. 14–30.

8. The National Parks Act of 1916, U.S. Statutes at Large, Vol. 39, p. 535.

9. Ibid., pp. 201–3.

10. Ronald A. Foresta, *America's National Parks and Their Keepers* (Washington, D.C.: Resources for the Future, 1984), pp. 18–30.

11. Ibid., p. 24.

12. Ibid., p. 95.

13. See ibid., p. 16; also see Susan Schrepfer, "Conflict in Preservation: The Sierra Club, Save-the-Redwoods League, and Redwood National Park," *Journal of Forest History* 24 (1980): 60–77.

14. Chase, *Playing God*, pp. 19–22.

15. Ibid., pp. 31–33.

16. A. Starker Leopold, et al., "Wildlife Management in the National Parks," Report of the Advisory Board on Wildlife Management to Secretary of Interior Stewart Udall, March 4, 1963.

17. Chase, *Playing God*, p. 34, based on Chase's interview with A. Starker Leopold.

18. Ibid., p. 35.

19. Darryll R. Johnson and James K. Agee, "Introduction to Ecosystem Management," in Agee and Johnson (eds.), *Ecosystem Management for Parks and Wilderness* (Seattle: University of Washington Press, 1988), p. 10. As Johnson and Agee note, official Park Service policy has never been revised to emphasize processes over vignettes.

20. Ibid., pp. 9ff. Also see N. L. Christensen, et al., "Review of Fire Management Program for Sequoia-Mixed Conifer Forests of Yosemite, Sequoia, and Kings Canyon National Parks," Report to Regional Director, National Park Service Western Region, San Francisco.

21. A variety of plans, embodying a variety of administrative structures, have been proposed. For a listing and discussion, see John D. Varley, "Managing Yellowstone National Park into the Twenty-first Century: The Park as Aquarium," in Agee and Johnson (eds.), *Ecosystem Management.* Also see McNamee, "Putting Nature First," pp. 14–15, for a brief discussion of political considerations and Paul Schullery, "Drawing the Lines in Yellowstone," *Orion Nature Quarterly,* Autumn (1986), for a cogent discussion of the problems of managing an ecosystem that is crisscrossed by political and administrative boundaries.

22. McNamee, "Putting Nature First," p. 8.

23. Johnson and Agee, "Introduction," p. 12.

24. Chase, *Playing God,* pp. 59–61.

25. Chase is particularly critical of Park Service scientists on this point, portraying them as "discounting sixty years of accumulated [scientific and eyewitness] evidence to the contrary" (p. 61), in order to proclaim the elk and bison to be merely at their historic maximum for the range. For a treatment more sympathetic to Park Service scientists and managers, see Schullery, "Drawing the Lines," p. 43.

   Fire policy and grizzly policy, especially regarding closing of the garbage dumps, are also areas of great controversy, too complex to discuss here. In each case, much controversy centers on whether the switch to natural regulation was based on adequate ecological and historical data.

26. R. H. MacArthur and E. O. Wilson, *The Theory of Island Biogeography* (Princeton: Princeton University Press, 1967).

27. Larry D. Harris, *The Fragmented Forest* (Chicago: University of Chicago Press, 1984), pp. 4–5.

28. Ibid., p. xvii.

29. Ibid., p. 107.

30. Ibid., pp. 89–90.

31. Ibid., p. 83.

32. Ibid., p. 156.

33. Ibid., pp. 141–44. It should be noted that Harris's idea of corridors is among the most controversial of his proposals. See, for example, Daniel Simberloff and James Cox, "Consequences and Costs of Conservation Corridors," *Conservation Biology* 1 (1987): 63–71.

34. For a brief summary of these early events, see Allen Krupnick, "The Chesapeake Bay Cleanup," *Resources,* Winter (1985): pp. 1–5. For a readable explanation of the biological processes involved, see Christopher D'Elia, "Nutrient Enrichment of the Chesapeake Bay: Too Much of a Good Thing," *Environment* 29 (March 1987).

35. See Allen Krupnick, "Reducing Bay Nutrients: An Economic Perspective," Resources for the Future Discussion Paper QE87-12, October 1987.

36. Tom Horton, "Remapping the Chesapeake," *The New American Land* (September/October 1987): 7–8.

37. See Tom Horton, *Bay Country* (Baltimore: Johns Hopkins University Press, 1987), p. 35.

38. Horton, "Remapping the Chesapeake," pp. 10, 13.

39. Ibid., p. 16.

40. Horton, *Bay Country*, p. xiii.

41. James M. Seif, Chairman of the Chesapeake Executive Council, "Foreword" to the *Second Annual Progress Report under the Chesapeake Bay Agreement* (Chesapeake Executive Council, February 1987).

42. Horton, "Remapping the Chesapeake," p. 13. See "Subtitle 15 Chesapeake Bay Critical Area Commission Criteria for Local Critical Area Program Development" issued under the authority of Natural Resources Article 8-1808(d), Annotated Code of Maryland, and State of Maryland, Chesapeake Bay Critical Area Commission, "A Guide to the Chesapeake Bay Critical Area Criteria," May 1986.

43. Krupnick, "Chesapeake Bay Cleanup," p. 5. Recently, strains have developed between developers and farmers, with the latter complaining that recent changes will shift the burden of limitations more heavily upon them. See *The Washington Post*, August 1989.

44. Horton, "Remapping the Chesapeake," p. 14.

45. See Eugene C. Hargrove, *Foundations of Environmental Ethics* (Englewood Cliffs, NJ: Prentice-Hall, 1989), chap. 5, for a discussion of therapeutic nihilism. While Hargrove's characterization of therapeutic nihilism is useful, I disagree with his suggestion that Leopold supported this theory, which is closely associated with what I have called isolationism.

46. Lewis Mumford, "Introduction" to Ian McHarg, *Design with Nature* (Garden City, NY: Doubleday/Natural History Press, 1969), p. viii.

47. Chase, *Playing God*, p. 353. Above, we noted that Chase attacks "whole ecosystem management" and "natural regulation" as if they are identical. Here, we are concerned with his attack on the former rather than the latter.

48. Ibid., pp. 316–17.

49. Ibid., p. 319.

50. Ibid., p. 325.

51. Ibid., chap. 19.

52. Ibid., p. 358.

53. Ibid., p. 361.

54. See J. B. Callicott, "Animal Liberation: A Triangular Affair," *Environmental Ethics* 2 (1980): 311–28; Tom Regan, *The Case for Animal Rights* (Berkeley: University of California Press, 1983), pp. 362–63; and J. Baird Callicott, *In Defense of the Land Ethic* (Albany: State University of New York Press, 1989), pp. 39–59, 91–94.

55. See Chase, *Playing God*, pp. 92–97, for a summary of the evidence supporting this conclusion.

56. John Craighead, "Yellowstone: The Bottom Line," *Orion Nature Quarterly*, Summer (1989): 569.

57. Alston Chase, "Greater Yellowstone and the Death and Rebirth of the National Parks Ideal," *Orion Nature Quarterly*, Summer (1989): 47.

58. Leopold, *Sand County*, pp. 176–77.
59. Chase, "Greater Yellowstone," p. 55.

Chapter 10

1. Wendell Berry, "Amplifications: Preserving Wilderness," *Wilderness* 50 (1987): 39–40, 50–54.
2. Bryan G. Norton, "Thoreau's Insect Analogies: Or, Why Environmentalists Hate Mainstream Economists," *Environmental Ethics* 13/3 (1991).
3. See Karen J. Warren, "Feminism and Ecology: Making Connections," *Environmental Ethics* 9 (Spring 1987).
4. Christopher Stone, *Earth and Other Ethics* (New York: Harper & Row, 1988), p. 116.
5. Ibid., p. 132.
6. Roderick Nash, *The Rights of Nature* (Madison: University of Wisconsin Press, 1989).
7. Nash, *The Rights of Nature*, pp. 38–40.
8. Ibid., p. 41. Nash also cites Bill Devall's criticism of Muir on similar grounds. "John Muir as Deep Ecologist," *Environmental Review* 6 (1982): 63–86. Nash notes that Muir, in this respect, began a trend, and he cites examples of Leopold, the Sierra Club, and species preservationists as having engaged in the same sort of political dishonesty.
9. Ibid., p. 76. Brief quotation from Krutch is from "Conservation Is Not Enough," *American Scholar* 24 (1954): 297.
10. See J. Baird Callicott, "The Case Against Moral Pluralism," *Environmental Ethics* 12 (1990): 99–124.
11. David Hume, *A Treatise of Human Nature* (London: Oxford University Press, 1965 printing), pp. 469–70.
12. See, for example, Paul Taylor, *Respect for Nature* (Princeton: Princeton University Press, 1986).

Chapter 11

1. Stephen Toulmin, *Human Understanding*, Vol. 1 (Princeton, NJ: Princeton University Press, 1972), p. 340.
2. Robert C. Mitchell, "From Conservation to Environmental Movement: The Development of the Modern Environmental Lobbies," Resources for the Future, Discussion Paper QE85-12, Washington, D.C., June, 1985.
3. See Talbot Page, "A Generic View of Toxic Chemicals and Similar Risks," *Ecology Law Quarterly* 7 (1978): 207–44; and Ezra Mishan and Talbot Page, "The Methodology of Cost Benefit Analysis—With particular Reference to the Ozone Problem," California Institute of Technology, Social Science Working Paper 249, January 1979.
4. See, for example, Herman B. Leonard and Richard J. Zeckhauser, "Cost-Benefit Analysis Defended" in Donald VanDeVeer and Christine Pierce, *People, Penguins, and Plastic Trees* (Belmont, CA: Wadsworth, 1986), pp. 249–53.
5. See Allen V. Kneese and William D. Schulze, "Ethics and Environmental Economics," in A. V. Kneese and J. L. Sweeney (eds.), *Handbook of Natural Resource and Energy Economics* (Amsterdam: North Holland, 1985), p. 191.
6. Mishan and Page, "Methodology," pp. 71–72.

7. See Guy Kirsch, Peter Nijkamp, and Klaus Zimmermann (eds.), *The Formulation of Time Preferences in a Multidisciplinary Perspective* (Aldershot, U.K.: Gower Publishing Company, 1988).

8. Kneese and Schulze, "Ethics," p 195; J. A. Doeleman, "On the Social Rate of Discount: The Case for Macroenvironmental Policy," *Environmental Ethics* 2 (Spring 1980).

9. Linda B. Brubaker, "Vegetation History and Anticipating Future Vegetation Change," in James K. Agee and Darryll R. Johnson (eds.), *Ecosystem Management for Parks and Wilderness* (Seattle: University of Washington Press, 1988), pp. 42–43.

10. Ibid., p. 49.

11. Ibid., p. 58.

12. John Passmore, *Man's Responsibility for Nature* (New York: Charles Scribner's Sons, 1974) p. 91. A less extreme form of this theory would recognize strong obligations to immediate posterity and somewhat weaker obligations to their successors. See, for example, Daniel Callahan, "What Obligations Do We Have to Future Generations?" *American Ecclesiastical Review* (1971): 265–80.

13. Admittedly, Passmore might reply that doing *nothing* about the greenhouse problem may entail some moral opprobrium, even on his theory, because this policy may impose on the next generation an unduly onerous task in the fulfillment of their duties to their children. But these indirect obligations seem ad hoc and, worse, negotiable: The indirect obligations to our children might be discharged by accumulating wealth that will be passed on to our children (and this will compensate them for the hardships they will bear to protect their children from calamity).

14. Passmore *Man's Responsibility* (p. 86) quotes the economist Joan Robinson to support his argument: "This problem [of sacrifics for posterity] cannot be resolved by any kind of calculation based on discounting the future, for the individuals concerned in the loss and gain are different. . . . The benefit from their sacrifices will come later and they may not survive to see it. The choice must be taken somehow or other, but the principles of Welfare Economics do not help to settle it." From Joan Robinson, *Economic Philosophy* (Harmondsworth, 1964), p. 115.

15. Ernest Partridge, "The Moral Uses of Future Generations," presented at the Annual Meeting of the Oregon Chapter of the American Fisheries Society, Bend, Oregon, February 8, 1989.

16. Traditional moral theorists, who assume that all obligations are obligations to individuals, and who have been unable to analyze this obligation as an aggregation of obligations to future individuals, have been puzzled by the widely felt obligation we feel to perpetuate the species. See, for example, Jan Narveson, "Utilitarianism and New Generations," *Mind* 76 (1967).

17. Edmund Burke, *Reflections on the Revolution in France* (London: Dent, 1910), pp. 93–94.

18. Aldo Leopold, *A Sand County Almanac* (London: Oxford University Press, 1949), pp. 200–201. One might ask, though it may seem impertinent to question Leopold's soaring prose, "Why not save wildness in books and movies?" And Leopold again has an eloquent answer: "Book-pigeons cannot dive out of a cloud to make the deer run for cover, or clap their wings in thunderous applause of mast-laden woods" (Ibid., p 109). See Hargrove, whose "ontological argument" for environmental ethics implies an obligation to protect natural objects be-

cause of their aesthetic value. Eugene C. Hargrove, *Foundations of Environmental Ethics* (Englewood Cliffs, NJ: Prentice-Hall, 1989), pp. 165–205.

Chapter 12

1. It must be noted that we cannot accept the deep ecologists' usual characterization of reform environmentalism. Devall and Sessions, for example, set their deep ecological position in opposition to "resource conservation and development," and illustrate this position with quotations from Pinchot, from a representative of the Industrial Forestry Association, and from an official of the Bureau of Reclamation under Reagan! But mainline environmental groups have, as we have seen above, repudiated Pinchot's ideas on resource use and have vehemently opposed production-oriented forestry and the Reagan approach to resource use generally.

   Devall and Sessions characterize the ethical bases of mainline environmentalism as resting on three beliefs: "(1) the metaphor of the market; (2) the Earth as a collection of human resources or potential commodities, and (3) the Earth as a machine." We have seen that mainline environmentalists have consistently and explicitly criticized and rejected (1)–(3). Bill Devall and George Sessions, "The Development of Natural Resources and the Integrity of Nature," *Environmental Ethics* 6 (1984): 294, 297, 301, 302.

2. Ibid., pp. 305–6.

3. Ibid., p. 312.

4. Ibid., p. 303.

5. See Paul Taylor, *Respect for Nature* (Princeton: Princeton University Press, 1986), for an attempt to work out more detailed "priority rules" for interspecific justice based on a strict analogy between human ethics of the Kantian variety and an environmental ethic. Taylor succeeds in justifying the building of museums, but only by violating his original purpose to maintain a strict parallel between human and environmental ethics. See Bryan G. Norton, "Review of Taylor's *Respect for Nature*," *Environmental Ethics* 9 (Fall 1987): 261–67.

6. Warwick Fox, "Deep Ecology: The New Philosophy of Our Time?" *The Ecologist* 14 (1984): 198–99.

7. At first this new thought experiment may seem hopeless. How can a "rational" chooser turn out to be a fish or a plant? Plants and fish do not, at first, recommend themselves as philosopher kings. But Rawls has already provided guidance in solving a similar problem that exists in the human-only case. Because Ric is, by hypothesis, ignorant of his future personal characteristics such as intelligence, mental health, and so on, he must design his society while keeping in mind that there is a small possibility that he will be severely retarded, mentally ill, or vegetative. While Ric is understood as a hypothetical chooser, he chooses *as if he were rational*. He receives no guarantees that he will remain so. We can deal similarly with cases in which Ric turns out in reality to be a cockroach or a Furbish lousewort. If Ric is sensitive to the possibility that he might be instantiated as a member of any species, he should design a society that will be an attractive habitation regardless of the accidents of instantiation.

8. Ric's choice corresponds to two alternative versions of justice among species. Extensionists, such as Tom Regan and Paul Taylor, interpret interspecific en-

vironmental justice as a requirement that all actions must take into account all legitimate interests of all relevant individuals. It is aptly called extensionism because it applies traditional, individualistic notions to more, perhaps all, species. Extensionism is an expression of individualism: All harms are harms to individuals. It is therefore monistic in the sense that, however broadly it applies the principles of individual-to-individual morality, it sees all moral obligations as deriving from effects upon, or obligations regarding, individuals. Contextualism, on the other hand, is pluralistic in the sense that it applies different criteria depending on contextual information, and in some contexts acts to protect the complexity and organization of systems even at the expense of individuals.

9. Frithjof Capra, *The Tao of Physics* (New York: Bantam Books, 1975), p. 69.
10. As J. Baird Callicott has perceptively pointed out, deep ecologists have gained considerable philosophical mileage from the longstanding ethical assumption, common to the Kantian, rights-based and the Benthamite, utilitarian traditions, that the individual self and its interests are intrinsically valuable and limited only by conflicting individual interests of equal or greater power. See Callicott, *In Defense of the Land Ethic* (Albany: State University of New York Press, 1989), pp. 172–73, 111–14.
11. Fox, "Deep Ecology, p. 198; based on Charles Birch and John Cobb, *The Liberation of Life: From Cell to the Community* (Cambridge: Cambridge University Press, 1981).
12. Fox, "Deep Ecology," p. 198.
13. Steve Sapontzis "Predation," *Ethics and Animals* 5 (June 1984): 27–36.
14. I realize that there are important ethical questions about *how* culling is carried out and how it affects animals individually. There may well be strong arguments for controlling herds of wild grazers only by birth control or sterilization, as we try to control human populations. Even so, three comments are in order: (1) this would *still* not make management practices "fair" in the sense we apply to human population control because the reproductive meddling could never, by the nature of elk and goat mental capacities, be voluntary, as we would require in the human case; (2) hunting has the advantage of continuing selective pressure; and (3) moral constraints guiding the *means* by which culling and removal of exotics is undertaken are properly questions in cross-specific, individualistic ethics, not environmental ethics.
15. See "The Aggregators and Moralists Revisited," in chap. 8, pp. 152–54.
16. Bill Devall and George Sessions, *Deep Ecology: Living As If Nature Mattered* (Salt Lake City: Peregrine Smith Books, 1985), p. 2.
17. Ibid., p. 4.
18. Alston Chase, "The Great, Green Deep Ecology Revolution," *Rolling Stone*, April 23, 1987, pp. 61–62.
19. Devall and Sessions, *Deep Ecology*, pp. 7–8.
20. Chase, "The Great Ecology Revolution," p. 62.
21. Ibid., p. 164.
22. Devall and Sessions, *Deep Ecology*, p. 8.
23. Ibid., p. 3.
24. Michael McCloskey, "The Environmental Movement from the Inside," discussion at meetings of the American Association for the Advancement of Science, New Orleans, February 18, 1990.

25. See Peter Borelli, "The Ecophilosophers," in Peter Borelli (ed.), *Crossroads: Environmental Priorities for the Future* (Washington, D.C.: Island Press, 1988), pp. 80–81.

26. Warwick Fox, "Approaching Deep Ecology: A Response to Richard Sylvan's Critique of Deep Ecology," Environmental Studies Occasional Paper #20, Centre for Environmental Studies, University of Tasmania, Hobart, Tasmania, 1986, p. 39.

27. Arne Naess, "Intuition, Intrinsic Value and Deep Ecology: A Reply to Fox," *The Ecologist* 14 (1984): 202.

28. Quoted in Fox, "Approaching Deep Ecology," p. 46.

29. J. Baird Callicott, "Intrinsic Value, Quantum Theory, and Environmental Ethics," in *In Defense of the Land Ethic*, chap. 9.

30. J. Baird Callicott, "On the Intrinsic Value of Nonhuman Species," in Bryan G. Norton (ed.), *The Preservation of Species* (Princeton: Princeton University Press, 1986), chap. 6.

31. Ibid., p. 133–34.

32. Interview with author, July 28, 1987.

33. For a more recent analogue, consider the economists' belief in the Axiom of Abundance. That axiom, which asserts that there exist suitable substitutes for every natural resource, cannot be disproven; it can always be said that we just have not found the best substitute as yet. This feature of certain beliefs—that they are empirical in form, but seem *impervious* to refutation—is common among the most basic axioms of a worldview.

34. Recognizing that the environmentalists' impulse is essentially a *systematic* impulse, we must still ask if there are also obligations that individual humans have to wild animals and plants. Is hunting immoral? Are leg-hold traps cruel and immoral? What are the obligations to individual animals that are displaced when humans drain a marsh for cultivation? Are obligations to mammals the same as obligations to amphibians? To plants? These questions form the proper domain of interspecific, individual ethics, an area currently being explored by animal liberationists. Environmentalists should, I think, be especially attentive to this debate because their activities and policies often affect individual wild animals, as when they "cull" a herd to fit its range or advocate controlled hunting to limit populations of birds or mammals. My point, rather, is that these questions of moral relations among individuals of different species are addressed on a different level of the moral hierarchy than are most characteristically environmental problems.

35. Mark Sagoff, "Animal Liberation and Environmental Ethics: Bad Marriage, Quick Divorce," *Osgoode Hall Law Journal* 22 (1984); Callicott, *In Defense of the Land Ethic*, p. 43.

Epilogue

1. Ed Bruske, "Anglers Wondering Why Bluefish Aren't Biting," *Washington Post*, July 18, 1989, p. B5.

2. Ibid.

3. See Chapter 9.

4. Garrett Hardin, "The Tragedy of the Commons," *Science*, 162 (1968) 1243–48.

5. In his original formulation, Hardin emphasized the common ownership feature of exploited commons: "Freedom in a commons brings ruin to all," (Ibid., p.

1244). Subsequently, economic analysis has shown that the common property case is a special case of a broader phenomenon and that, under not unlikely circumstances, individual economic incentives can encourage owners to exploit to destruction even resources in which they have a proprietary interest. See Colin Clark, "The Economics of Overexploitation," *Science* 181 (1974): 630–34: Colin Clark, *Mathematical Bioeconomics* (New York: John Wiley, 1976); and Daniel Fife, "Killing the Goose," *Environment* 13 (1971): 20–27.

6. T. F. H. Allen and Thomas B. Starr, *Hierarchy: Perspective for Ecological Complexity* (Chicago: University of Chicago Press, 1982), pp. 19–30.

7. Interview with author, April 15, 1990.

8. Annie Dillard, *Pilgrim at Tinker Creek* (New York: Harper and Row, 1974), p. 175.

Index